GENE EXPRESSION AND CELL–CELL INTERACTIONS IN THE DEVELOPING NERVOUS SYSTEM

ADVANCES IN EXPERIMENTAL MEDICINE AND BIOLOGY

Recent Volumes in this Series

GENE EXPRESSION AND CELL–CELL INTERACTIONS IN THE DEVELOPING NERVOUS SYSTEM

Edited by

Jean M. Lauder

University of North Carolina School of Medicine
Chapel Hill, North Carolina

and

Phillip G. Nelson

National Institutes of Health
Bethesda, Maryland

PLENUM PRESS • NEW YORK AND LONDON

Library of Congress Cataloging in Publication Data

International Society for Developmental Neuroscience Congress (4th: 1983: Salt Lake City, Utah)
Gene expression and cell-cell interactions in the developing nervous systems.

(Advances in experimental medicine and biology; v. 181)
"Proceedings of selected symposia from the Fourth Congress of the International Society for Developmental Neuroscience, held July 3–7, 1983, in Salt Lake City, Utah" —Verso of t.p.
Includes bibliographical references and index.
1. Developmental neurology — Congresses. 2. Gene expression — Congresses. 3. Cell interaction — Congresses. 4. Neurogenetics — Congresses. I. Lauder, Jean M. II. Nelson, Phillip Gillard, 1931–   . III. Title. IV. Series.
QP363.5.I566   1983   612′.8   84-22892
ISBN 0-306-41836-3

Proceedings of selected symposia from the Fourth
Congress of the International Society for
Developmental Neuroscience, held July 3–7, 1983,
in Salt Lake City, Utah

©1984 Plenum Press, New York
A Division of Plenum Publishing Corporation
233 Spring Street, New York, N.Y. 10013

Printed in the United States of America

PREFACE

The dramatic advances in molecular genetics are becoming incorporated into neurobiologic studies at an ever increasing rate. In developmental neurobiology, the importance of cell-cell interactions for neurogenesis and gene expression is beginning to be understood in terms of the molecular bases for these interactions. This book seeks to emphasize the importance of molecular technology in the study of neurogenetic mechanisms and to explore the possible relationships between specific cell-cell interactions and regulated gene expression in the developing nervous system.

This volume consists of nineteen chapters which address questions of gene expression and the importance of cell-cell interactions as key factors in the developing nervous system. Rather than viewing these two processes as separate mechanisms, as the organization of these chapters might suggest, we would like to emphasize the interplay of these genetic and epigenetic influences in all phases of neural ontogeny, a concept which is made clear by the subject matter of the contributions themselves. The authors of these chapters were participants in selected symposia from the Fourth Congress of the International Society of Developmental Neuroscience held in Salt Lake City, Utah, July 3-7, 1983.

The first two papers, under the heading "Perspectives", were written by the two plenary speakers for this meeting who were asked to provide a theoretical overview of their work, based on the content of their original lectures. The chapter by Karl Pfenninger presents a model of the molecular events involved in growth cone motility and chemotaxis, based on his recent work with isolated growth cone fractions. In the chapter by Marshall Nirenberg, the growth cone is followed to the synapse where the importance of cAMP as a regulator of synaptogenesis is discussed.

In the section on "Molecular Genetics and Gene Expression" we have collected papers which range from the use of recombinant DNA technology to probe the molecular basis for development of cellular diversity in the nervous system to the use of mutants to study mechanisms of circuitry construction in the developing brain.

In the chapter by <u>Devillers-Thiery</u> and co-workers, the molecular cloning of the acetylcholine receptor is described, including their hypothesis of the transmembrane organization of this molecule. This chapter illustrates the power of biotechnology for the elucidation of molecular mechanisms of receptor development and provides a context for the following chapters which make use of this technology for studies of genetic mechanisms at play in neurogenesis.

In the following contribution by <u>Hahn and Chaudhari</u>, the macromolecular complexity of both the developing and adult brain is discussed, including the importance of both genetic and epigenetic mechanisms in creating this diversity. These studies employ the strategy of making recombinant DNA (cDNA) libraries from adult brain and screening them by hybridization against brain messenger RNA (mRNA) from various stages of development to detect developmentally regulated genes and gene products.

<u>Joh</u>, on the other hand, discusses the molecular biology of a specific class of brain-specific molecules, those enzymes involved in the synthesis of the catecholamine neurotransmitters. In this contribution, a case is made for homology between the genes coding for these enzymes and for their derivation from a common ancestral gene. He also postulates that these enzymes share common protein domains with nerve ending proteins with which they may be intimately associated either in the cell membrane or in the membranes of synaptic vesicles. He concludes that the coordinate expression of the genes coding for these enzymes may be important in determining the catecholaminergic phenotype.

In the chapter by <u>Pintar</u>, another class of neurotransmitters is discussed, the pro-opiomelanocortin (POMC) peptides. This chapter describes studies designed to ascertain mechanisms of differentiation of the pituitary using both classical biochemical approaches as well as recombinant DNA technology to determine the timing and regulation of POMC gene expression. The appearance of monoamine oxidase is also studied as another transmitter-related marker of pituitary differentiation.

The interplay of genetic and epigenetic influences in brain development is made especially clear in the chapter by <u>Villa-Komaroff</u> and colleagues who discuss the molecular biology of the insulin-like growth factors which appear to be localized in the developing brain and may play important roles in the regulation of its development. This chapter will be especially useful to those neurobiologists new to recombinant DNA technology, since the authors give a detailed and well organized description of the methodology they have used, including the rationale behind the various techniques chosen.

The chapter by <u>Dreyer and Roman</u> concludes the portion of this section devoted to the use of biotechnology by discussing an hypothesis for gene expression during embryogenesis based on the molecular biology of the immune system, in which the editing of chromosomes by the splicing of genes plays a key role in the regulated expression of selected gene sequences leading to the formation of specific cell lineages.

With the chapter by <u>Crepel and colleagues</u>, we return to the organismal level for a discussion of how mutations of the nervous system can be used to study mechanisms of synaptogenesis, using the cerebellum as an example. The model system is the multiple innervation of the Purkinje cells by climbing fibers which regresses to innervation by single fibers during postnatal development. In three different mutants (weaver, reeler, and staggerer) in which other key competitive elements are absent, this synaptic rearrangement does not occur. Thus the case is made for the importance of synaptic competition in the correct wiring of this brain circuitry.

The use of mutants to study developmental mechanisms in the nervous system is further explored in the chapter by <u>Wolf and Billings-Gagliardi</u> who describe their studies of three mutants which exhibit disorders of myelination. The locus of the mutation in all three cases appears to lie in the oligodendrocyte rather than in the developing axon, as determined from tissue culture studies where mutant and normal neurons and glia are mixed and the consequences for myelin formation analyzed.

The importance of ploidy for development of normal neuronal connections is discussed by <u>Tompkins and colleagues</u> who have studied the effects of polyploidy in amphibians on the development of the retino-tectal system. They postulate that alterations in cell size and number as a result of gene duplication can lead to a complex miswiring of the nervous system due to changes in cell-cell interactions. This chapter provides another example of the interplay of genetic and epigenetic mechanisms in neurogenesis, as discussed in detail in the next group of chapters.

The second section of this book contains a series of chapters which discuss various types of <u>"Cell-Cell Interactions and Epigenetic Influences"</u> which appear to play roles in shaping the developing nervous system, including possible molecular mechanisms involved in these processes.

This section begins with two chapters on cell-adhesion and recognition molecules which appear to be important in mediating these cell-cell interactions.

In the chapter by <u>Hoffman and Edelman</u>, a mechanism of cell adhesion is discussed based on the binding of cell adhesion mole-

cules (CAMs) on one cell to those on another adjacent cell. It is
suggested that a small number of these cell adhesion molecules
could mediate pattern formation in the developing nervous system
by undergoing chemical modification or changes in the spatio-
temporal patterns of their expression as a result of modulation
through such cell-cell interactions or other epigenetic influences.

<u>Marchase</u> discusses another cell surface molecule, ligatin,
which appears to be important in mediating the adhesion of retinal
cells to each other and to tectal cells via glycoproteins attached
to the ligatin molecule which is embedded in the cell membrane.
Such a mechanism may also be involved in cell-cell interactions in
other parts of the developing nervous system. The trafficking of
the ligatin molecule and its associated glycoproteins from their
sites of synthesis within the cell to their final destinations at
the cell surface is discussed in detail based on recent experimental findings.

New instrumentation to facilitate the microchemical analysis of
cell surface molecules is described in the chapter by <u>Dreyer and
colleagues</u> who have developed an ultrasensitive protein sequenator
and a miniaturized mass spectrometer which make possible the structural analysis of picomole quantities of proteins and peptides.
These peptide sequences are then used to synthesize oligonucleo-
tides which are used to probe cDNA or genomic libraries for the
genes related to these cell surface molecules. This instrumentation, together with advanced computer technology, is being used to
analyze molecules important in cell-cell interactions in various
parts of the developing nervous system, including the cerebellum
and the neural retina.

The functional consequences of specific cell-cell interactions
in the deveoping nervous system are discussed in the chapters by
<u>Linser and Moscona</u> and <u>Fisher</u> who correlate specific neuronal-glial
interactions with the development of glial marker enzymes in the
neural retina and cerebellum, respectively. In the neural retina,
the glial enzymes glutamine synthetase and carbonic anhydrase are
regulated differently, such that neuronal-glial interactions are
required for normal expression of glutamine synthetase, but not
for carbonic anhydrase. In cerebellar mutants, changes in the
expression of the glial marker, glycerol-3-phosphate dehydrogenase
(GPDH) are attributed to alterations in cell-cell interactions
between Bergmann glia and Purkinje cell dendrites.

The possible role of neurons and their neurotransmitters in
the development of their target cells is discussed in the chapter
by <u>Wolff and co-workers</u> with respect to the GABAergic system of the
developing visual cortex, based on the presence of the cellular
machinery for synthesis, uptake and release of GABA prior to the
formation of GABAergic synapses with other cortical cells.

In the last two chapters on synaptogenesis, <u>Fishman</u> and

Fishman and Nelson discuss the different phases which developing synapses pass through during synaptogenesis: targeting, stabilization and rearrangment. This sequence is similar to that described by Devillers-Thiery and colleagues in their discussion of the formation of the cholinergic synapse in the first chapter of this volume. The progression of developing synapses through these phases appears to be dependent on competition, as also suggested by Crepel in his chapter on cerebellar synaptogenesis in mutant mice, and is related to neuronal activity, as determined in cell culture experiments. The molecular basis for this activity-dependence may involve transported proteins or other, smaller molecules, such as peptides or ions.

The editors wish to thank all of the authors for their contributions to this volume, the International Society for Developmental Neuroscience for the hosting of these Symposia and Plenary lectures, and especially Mr. Philip Alvarez of Plenum Publishing Corporation for his own initiation of this book and his efficient handling of its production.

Jean Lauder
Phillip Nelson

CONTENTS

MOLECULAR BIOLOGY OF THE NERVE GROWTH CONE: A PERSPECTIVE

Karl H. Pfenninger

Dept. of Anatomy & Cell Biology
Columbia University
New York, N.Y.   10032

"I had the good fortune to behold for the first
time that fantastic ending of the growing
axon. In my sections of the three-days chick
embryo, this ending appeared as a concentration
of protoplasm of conical form, endowed with
amoeboid movements. It could be compared to a
living battering-ram, soft and flexible, which
advances, pushing aside mechanically the
obstacles which it finds in its way, until it
reaches the area of its peripheral distribution.
This curious terminal club, I christened the
growth cone" (Ramón y Cajal, 1937).

INTRODUCTION

As so vividly described by Cajal, the nerve growth cone is
the amoeboid, leading edge of the growing neurite. It occurs
only during a narrow segment of the neuron's life, during the
time between terminal mitosis and synaptogenesis and, possibly in
somewhat different form, during regeneration. As the growing,
advancing tip, the nerve growth cone plays a crucial role in
nervous system development. It is critical for the establishment
of the neuron's unusual geometry and high degree of polarity
because it appears to be the main site of insertion of new plasma-

lemmal components for surface expansion (Pfenninger and Maylié-
Pfenninger, 1981b; Pfenninger and Johnson, 1983). Furthermore, the
growth cone is the structure responsible for vectorial growth into
the appropriate target area because it is capable of locomotion,
pathfinding and chemotaxis (see below). Last but not least, it is
the structure which recognizes the target cell and triggers synaptogenesis. Nerve growth cones, devoid of ribosomes and Golgi
apparatus, are apparently not capable of performing many of the
synthetic functions carried out in the perikaryon (e.g., protein
synthesis), but they seem to be competent to execute a variety of
complex short-term functions, such as chemotaxis, without the
direct involvement of the perikaryon. Indeed, the time frame of
the growth cone's motile behavior (minutes) is so short that
constant signaling between growth cone and perikaryon for the
control of chemotaxis appears highly unlikely. Thus, the nerve
growth cone is endowed with a certain degree of autonomy, especially with regard to the control of vectorial growth: One may actually envision the nerve growth cone as a "leukocyte on a leash",
spinning out the neurite behind it (Figure 1). This analogy may be
useful for the formulation of hypotheses on growth cone function
and for the design of appropriate experimentation to test them.

Progress in research on nervous system development will
depend to a large extent on the identification of growth-cone-
specific components expected to be involved in functions unique
to the structure and on the elucidation of mechanisms controlling vectorial growth and triggering synaptogenesis. Any research
in these directions necessitates the use of methods of modern cell
and molecular biology and, thus, the availability of substantial
amounts of growth cone material in relatively pure form. For this
reason, my laboratory has invested several years of intense effort

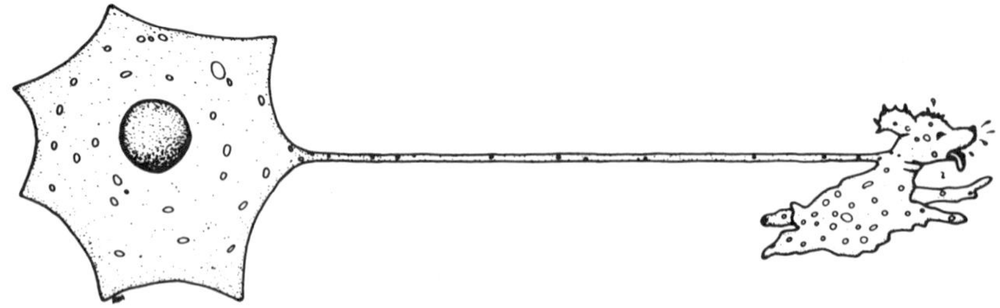

Fig. 1.   The nerve growth cone, a "leukocyte on a leash".

in developing methods for the isolation of nerve growth cones from fetal mammalian brain. We have recently succeeded with this endeavor (Pfenninger et al., 1983) and many of the observations and conclusions summarized here are based on new results obtained from this growth cone fraction.

RESULTS AND DISCUSSION

Because the subcellular fraction enriched in fragments of nerve growth cones, so-called "growth cone particles", plays a critical role in the studies discussed below, it is briefly described here: The homogenate of fetal rat brains (17 days gestation) is first spun to produce a low-speed supernatant. This supernatant is layered on a discontinuous sucrose density gradient and spun to equilibrium (Pfenninger et al., 1983). The fraction at the interface between the load (0.32 M sucrose) and 0.75 M sucrose is highly homogeneous and contains membrane-bound elements that bear the morphological characteristics of axonal nerve growth cones (Figure 2). These particles also co-purify with identified nerve growth cones microdissected from

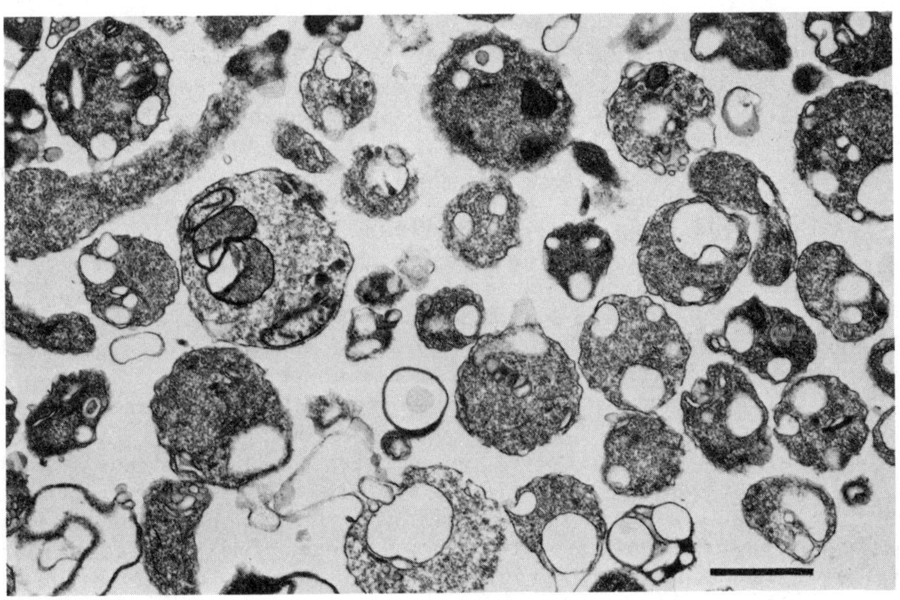

Fig. 2. Growth cone particles isolated by subcellular fractionation from fetal rat brain. The fraction is highly homo-
genous and consists of membrane-bound particles filled
with organelles characteristic of nerve growth cones. For
further description, see text. Calibration, 1μm.

primary cultures. A number of biochemical markers have recently been identified, and they support the cytological findings as follows: (i) One of the major membrane proteins of growth cone particles seems to have the same electrophoretic properties as GAP 43, a "growth-associated protein" which is neuron-specific and expressed during growth and regeneration, but not in mature axons (e.g., Skene and Willard, 1981; Ellis and Pfenninger, 1984; Meiri, Katz, Ellis, Willard, Pfenninger, in progress). (ii) Growth cone membranes are highly enriched in a large glycoprotein (250 to 300 kd) which is present during axonal growth, but neither in the dividing neuroblasts nor in the mature neuron (Wallis et al., 1983). (iii) The synaptic terminal's characteristic pattern of protein kinases and substrates is present in growth cone parti- cles and highly enriched in this fraction relative to fetal brain homogenate (Ellis et al., 1983; Ellis et al., 1984; Katz et al., 1984). The presence of these neuronal markers, together with the ultrastructural features of growth cone particles, is strong evidence for their origin from nerve growth cones. Therefore, biochemical analyses of growth cone particles can be used to investigate nerve growth cone properties and functions.

Growth-Specific Gene Products of the Neuron

The fact that the nerve growth cone is a unique structure reserved for vectorial growth of the neurite has long suggested that the neuron should be synthesizing specific components during sprouting. This view is further supported by the fact that the growth cone membrane has unusual properties, both from the stand- point of lectin labeling (Pfenninger and Maylié-Pfenninger, 1981a) and of intramembranous structure revealed by freeze-fracture (Pfenninger and Bunge, 1974; Small and Pfenninger, 1984). Howev- er, biochemical identification of the first growth-associated components was accomplished only recently by Hall et al. (1978), Theiler and McClure (1978), Bisby (1980), Benowitz et al. (1981), Heacock and Agranoff (1982), and by Skene and Willard (e.g., 1981), who named them "GAPs". GAPs are either components of membranes or, possibly, contained in the lumen of vesicular structures shuttled by rapid transport to the tip of the growing axon. As already mentioned, a collaborative effort between the Willard and Pfennin- ger laboratories has indeed suggested that GAP 43 co-migrates in two-dimensional gel electrophoresis with one of the major bona-fide membrane proteins of isolated growth cones (Meiri, Katz, Ellis, Willard, Pfenninger, in progress). It appears that, in our gel system, this protein forms a doublet with an apparent molecular weight of 38,000. A second major membrane protein of the growth cone, which migrates as a doublet at 34 kd, has not yet been characterized. The third of these membrane proteins migrates at 46 kd and has an acidic isoelectric point (4.3). It is now known to be a major phosphoprotein of the growth cone and will be discussed further below. All three proteins are highly enriched in growth cone membranes and, thus, are potential growth cone markers (Ellis and Pfenninger, 1984).

A different approach for the identification of growth-specific components of axons is the immunochemical comparison of nerve growth cone fragments with a well-defined neuronal preparation from a different stage of differentiation, e.g., synaptosomes. This can be achieved by generating monoclonal antibodies to growth cone membranes and selecting only those antibodies which recognize components of growth cones but not those of synaptosomes. One such antibody, named 5B4, recognizes in fetal brain and nerve growth cones a large membrane glycoprotein (Wallis et al., 1983). This antigen is neuraminidase-sensitive and forms a very broad band between about 250 and 300 kd in SDS-polyacrylamide gels. It is developmentally regulated, occurring mainly during the time of maximal growth of the brain (between about 17 days gestation and 8 days postnatal in the rat). Immunocytochemical studies show that the antigen's distribution is co-extensive with developing fiber tracts in the central nervous system. The antigen is not confined to the growth cone region but seems to be present along the shaft of the growing neurite as well. Interestingly, this growth-specific antigen is replaced in the mature brain by a smaller and sparser molecule (approximately 180 kd), which cross-reacts with the same antibody. This situation is somewhat reminiscent of N-CAM described by Edelman and collaborators (e.g., Hoffman et al., 1982) but, for various reasons, we believe that N-CAM is a different molecule.

Protein kinases and their substrates are well known for presynaptic nerve terminals, and some of them (e.g., synapsin I and its kinase) are among the most specific known biochemical markers of neuron and synapse. Because of the precursor-product relationship between growth cones and synaptic terminals, it is of substantial interest to search for phosphoproteins in nerve growth cones. The results of this analysis are rather surprising (Ellis et al., 1983, 1984; Katz et al., 1984). Although the growth cone particle fraction is devoid of synaptosomes and contains only a very small number of synaptic vesicles (Pfenninger et al., 1983) it contains all of the major cAMP- and calcium-dependent protein kinase activities of the synaptic terminal. However, the phosphorylation patterns are quantitatively very different. While synapsin I and its kinase are the predominant phosphoproteins of the synaptosome, they seem to be relatively minor substrates in the growth cone. By contrast, an acidic 46 kd protein, phosphorylated by a calcium/calmodulin-dependent kinase, and two acidic proteins of 40 and 80 kd, substrates of a calcium/ phospholipid-dependent kinase, are prevalent in nerve growth cones but minor in synaptosomes. It is not clear at present whether these differences between growth cones and nerve terminals are due to differences in the amounts of kinase, substrate, co-factors, or a combination of these parameters. Nevertheless, one can conclude that these kinase activities are developmentally regulated. It is of particular interest that the 46 kd substrate (phosphorylated in calcium/calmodulin-dependent fashion) is electrophoretically identical with the major 46 kd membrane protein of the growth cone. Overall, it is evident that the phosphorylation of certain protein kinase substrates is characteristically enhanced during neuritic growth. The potential relationship of these kinase activities with control mechanisms of neuritic growth is discussed below.

Control Mechanisms of Neuritic Growth

    Vectorial growth of the neurite involves a series of cellular mechanisms which include: (a) plasmalemmal expansion, which is believed to occur by exocytosis of precursor vesicles in the growth cone region (Pfenninger and Maylié-Pfenninger, 1981b; Pfenninger and Johnson, 1983; cf. Bray 1970, 1973); (b) cytoskeletal growth; (c) orientation and re-orientation of the growth cone by localized cytoskeletal rearrangements; (d) attachment and detachment of the growth cone to and from the substratum. Our knowledge of the mechanisms controlling these functions is very limited at present and almost entirely confined to the action of nerve growth factor (NGF) on neurons of the peripheral nervous system. NGF has long been known to enhance neuronal survival and to stimulate neuritic growth on a long-term basis (for review, see Greene and Shooter, 1980). Recently, Gundersen and Barrett (1980) discovered that NGF also elicits a short-term response: In addition to its trophic effect, NGF acts as a chemotactic factor on growth cones (see also Letourneau, 1978). Another critical discovery was made by Campenot (1977), who found that NGF must be present in the growth cone's microenvironment to be able to exert its effect; NGF does not stimulate neuronal growth if it is present only near the perikaryon. This suggests that "functional" NGF receptors are concentrated in the growth cone region. The NGF receptor is known to be internalized by receptor-mediated endocytosis, and NGF is known to be retrogradely transported to the perikaryon (Hendry et al., 1974). However, it is not known whether this phenomenon reflects receptor down-regulation under the experimental conditions used, or whether it is related to the transduction of the NGF signal to the cell's perikaryon. Many studies have been performed in attempts to elucidate the receptor-activated mechanisms stimulated by NGF, but most of this work has remained inconclusive (for review, see Greene and Shooter, 1980).

A number of cell surface ligands are now known to stimulate phospholipid transmethylation, a process by which phosphatidylethanolamine is converted into phosphatidylcholine and transferred from the inner to the outer membrane leaflet (for review, see Hirata and Axelrod, 1980). Phospholipid transmethylation is also believed to trigger calcium influx into the cell and, ultimately, to stimulate phospholipase A_2, probably via the activation of protein kinases. The activation of this phospholipase and cleavage of phosphatidylcholine to lysophosphatidylcholine and arachidonic acid have been proposed as mechanisms leading to exocytotic membrane fusion. The best studied systems in which this type of cascade seems to operate are mast cells and mast-cell-like leukemia cells, where the clustering of IgE receptors eventually triggers exocytotic histamine release (Crews et al., 1981; McGivney et al., 1981). The activation of phospholipase A_2 also seems to be an important event in leukocyte chemotaxis (e.g., Snyderman and Goetzl, 1981). These findings have prompted us to investigate phospholipid transmethylation in nerve growth cones. Indeed, NGF was found to lead to very rapid and transient phospholipid methylation in the distal neurites of sympathetic neurons grown in culture, but not in the more proximal neurites and perikarya (Pfenninger and Johnson, 1981). Interestingly, blockers of phospholipid transmethylation have recently been found to inhibit NGF responses in PC 12 cells (Seeley et al., 1984). Our transmethylation studies also revealed the generation of lysophosphatidylcholine in distal neurites. This latter observation suggests the presence of phospholipase A_2 near the methylation site. More recently, we have assayed phospholipid transmethylase activity in growth cone particles. Such measurements have already clearly established the presence of phospholipid transmethylases in growth cones of the CNS (Garofalo and Pfenninger, in preparation).

NGF-activated phospholipid transmethylation in growth cones of peripheral neurons is consistent with the idea that a transducer cascade similar to the one proposed by Hirata and Axelrod (1980) may be the mechanism by which it activates the initial intracellular events (see Figure 3). However, the evidence for this hypothesis is far from being complete, and the only statements that can be made with confidence at the present time are (a) that phospholipid transmethylation is the earliest known biochemical event triggered by NGF, and (b) that this event is confined to that domain of the neuron which contains the "competent" receptor (Campenot, 1977; cf. above). If phospholipid transmethylation does indeed trigger calcium influx, as Hirata and Axelrod propose, then this is consistent with the observation that a calcium gradient, applied to nerve growth cones in the presence of calcium ionophore, elicits a chemotactic response similar to that seen with NGF (Gundersen and Barrett, 1980). Also of interest is the fact that growth cones are believed to be rich in calcium channels (e.g., Llinas and Sugimori, 1979; Spitzer, 1979).

Localized flux of calcium into the nerve growth cone can easily be thought of as regionally altering cytoskeletal configuration and, thus, leading to growth cone re-orientation (see Figure 3). Furthermore, changes in intracellular calcium levels are certain to affect protein kinase activities. As already pointed out, such activities are quite prominent in growth cones isolated from fetal brain. The predominant substrates include a major membrane protein, pp46, which is phosphorylated in calcium/calmodulin-dependent manner, and two acidic proteins of 40 and 80 kd, which are phosphorylated by a calcium/phospholipid-dependent kinase, probably protein kinase C (Katz et al., 1984). Because phosphorylation of these substrates is greatly enhanced in growth cone particles compared to fetal brain homogenate, low-speed pellet and low-speed supernatant, and because these substrates are far less prominent in synaptic nerve terminals, these kinase activities may well be involved in growth-cone-specific regulatory mechanisms. The involvement of (putative) protein kinase C is particularly intriguing because this kinase is believed to be the receptor for phorbol esters (e.g., Niedel et al., 1983), a group of potent tumor-promoting agents that are known to influence cellular differentiation and growth as well as leukocyte chemotaxis (e.g., Weinstein, 1981; Laskin et al., 1981). In addition, the generation of lysophosphatidylcholine by phospholipase A_2 may lead to fusion of plasmalemmal precursor vesicles with the cell surface (cf. Poole et al., 1970) and, thus, to the membrane expansion necessary for neuritic growth (see Figure 3). A metabolite of the arachidonic acid also liberated in the process could serve as a messenger to the perikaryon for transmitting the growth factor's long-term trophic signal.

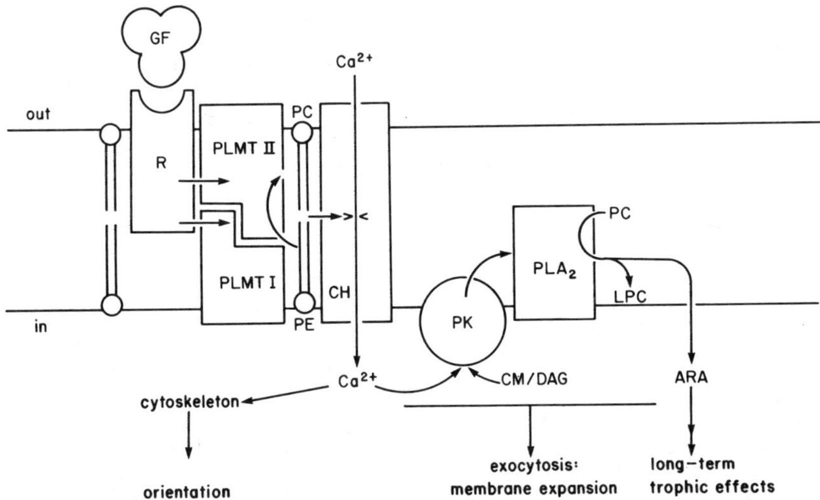

Fig. 3. Hypothetical model of the transducer mechanism by which a growth factor may stimulate vectorial growth of the neurite. The proposed mechanism consists of an enzyme cascade initiated by receptor-activated phospholipid transmethylation. Abbreviations: ARA, arachidonic acid; CH, Ca^{2+} channel; CM, calmodulin; DAG, diacylglycerol; GF, growth factor; LPC, lysophosphatidylcholine; PC, phosphatidylcholine; PE, phosphatidylethanolamine; PK, protein kinase(s); PLA_2, phospholipase A_2; PLMT I, II, phospholipid methyltransferases; R, receptor.

Molecular Changes During Synaptogenesis

The distinctive morphologies and functional properties of growth cones and synaptic terminals suggest that the two structures each contain a set of specific "marker" proteins. Such molecules should be revealed by the biochemical comparison of fractions enriched in the respective structures. However, a simple biochemical comparison would mainly stress differences due to the presence in synaptosomes (but not growth cone particles) of postsynaptic densities. This problem can be overcome by studying axoplasmic transport of newly synthesized, radiolabeled components into growing and mature axonal endings. This approach is similar to that used to identify GAPs (see above) but includes the isolation of the specific terminal structures as follows: The eyes of fetal (16 days gestation) and adult rats are injected with ^{35}S-methionine: after appropriate survival times to allow for newly synthesized proteins to be transported into the axon periphery, growth cone particles and synaptosomes are isolated from the forebrains of the fetal and adult animals, respectively. Radiolabeled proteins are then analyzed by SDS-polyacrylamide gel electrophoresis and radioautography. Juxtaposition of the radioautograms reveals major differences (Simkowitz and Pfenninger, 1983): The growth cone fraction contains a series of major radioactive bands including actin, tubulin and the major membrane proteins of the growth cone (46, 38, 34 kd). The 38 and 34 kd proteins do not have labeled counterparts in synaptic terminals. In contrast, there is only one major band, migrating at approximately 27 kd, in the synaptosome fraction. This polypeptide does not appear to have a counterpart in the growth cone preparation. Some of the less striking synaptic bands have corresponding species in the growth cones, whereas others seem to be unique to synaptic terminals. These experiments demonstrate dramatic differences in the rates of synthesis of specific proteins of growth cones versus synaptic endings and suggest that a quite different set of genes is expressed in the synapsing, mature neuron compared to its sprouting precursor.

This view is further confirmed by the significant shifts in protein kinase activities from the growth cone to the presynaptic ending (cf. above) and by the results of immunochemical comparisons carried out in this laboratory. In a manner analogous to the one described earlier, synaptosomes can be used to generate monoclonal antibodies. Screening of these antibodies to eliminate those which cross-react with growth cone particles can then be used to obtain synaptosome-specific probes. One such antibody, termed 2L4, recognizes a plasmalemmal component of the presynaptic ending that is protease-, SDS- and aldehyde-sensitive and, therefore, most likely a protein (Wallis et al., 1983). This then is a molecule whose expression is likely to be triggered by synaptogenesis.

CONCLUSIONS

The successful isolation of nerve growth cone fragments in bulk is likely to have major impact on our understanding of

cellular and molecular mechanisms in nervous system development
because it provides the tool for a series of critical and hereto-
fore impossible biochemical analyses. I have proposed an enzyme
cascade that may be involved in transduction of growth factor
signals for neuritic growth and chemotaxis. This hypothesis can
now be tested with biochemical methods. Biochemical comparison of
growth cones with presynaptic terminals has already led to the
identification of a number of putative differentiation markers of
the neuron. The further study of such molecules in a biological
context is likely to yield new information on growth-cone- and
synapse-specific functions. In addition, probes for such molecules
can serve to monitor nerve cell differentiation biochemically and,
thus, to study mechanisms that regulate neuronal growth and synaptogenesis, as well as sprouting and regeneration in the mature
nervous system. Given the technological advances in molecular
biology, such studies can be carried through to the DNA level
relatively easily once appropriate gene products have been identified.

ACKNOWLEDGEMENTS

I would like to thank the past and present members of my
laboratory for their dedication and contribution to this work. I
am particularly indebted to Drs. Leland C. Ellis, Robert S.
Garofalo, Carolyn Hyman, Flora Katz, Stefan Somlo, Ira Wallis, and
to Edith Abreu, Linda B. Friedman, Marian P. Johnson and Philip
Simkowitz.

This work was supported by grants from the National Institutes of Health (NINCDS), the National Science Foundation, the
National Spinal Cord Injury Foundation, and the Henry G. Stifel
III Spinal Cord Injury Foundation. Fellowship support was provided
by the National Institutes of Health and the Muscular Dystrophy
Association of America. Karl H. Pfenninger also was the recipient
of an Irma T. Hirschl Career Scientist Award.

REFERENCES

Benowitz, L.I., Shashoua, V.E., and Yoon, M.G., 1981, Specific
        changes in rapidly transported proteins during regenera-
        tion of the goldfish optic nerve, J. Neurosci., 1:300-307.
Bisby, M.A., 1980, Changes in the composition of labeled protein
        transported in motor axons during their regeneration, J.
        Neurobiol., 11:435-446.

Bray, D., 1970, Surface movements during the growth of single
    explanted neurons, Proc. Natl. Acad. Sci. USA, 65:905-910.
Bray, D., 1973, Branching patterns of isolated sympathetic
    neurons, J. Cell Biol., 56:702-712.
Campenot, R.B., 1977, Local control of neurite development by
    nerve growth factor, Proc. Natl. Acad. Sci. USA, 74:4516-
    4519.
Crews, F.T., Morita, Y., McGivney, A., Hirata, F., Siraganian,
    R.P., and Axelrod, J., 1981, IgE mediated histamine
    release in rat basophilic leukemia cells. Receptor
    activation, phospholipid methylation, Ca^{2+} flux and
    release of arachidonic acid, Arch. Biochem. Biophys.,
    212:561-571.
Ellis, L., Katz, R., and Pfenninger, K.H., 1983, Characterization
    of phosphoproteins of isolated nerve growth cone parti-
    cles, J. Cell Biol., 97:242a.
Ellis, L., Katz, F., and Pfenninger, K.H., 1984, Nerve growth
    cones isolated from fetal rat brain: III. cAMP-binding
    proteins and cAMP-dependent protein phosphorylation,
    in submission.
Ellis, L., and Pfenninger, K.H., 1984, Nerve growth cones isolat-
    ed from fetal rat brain: II. Polypeptides of growth cone
    membranes, in submission.
Greene, L.A., and Shooter, E.M., 1980, The nerve growth factor:
    Biochemistry, synthesis and mechanisms of action, Ann.
    Rev. Neurosci., 3:353-402.
Gundersen, R.W., and Barrett, J.N., 1980, Characterization of
    the turning response of dorsal root neurites toward nerve
    growth factor, J. Cell. Biol., 87:546-554.
Hall, M.E., Wilson, D.L., and Stone, G.C., 1978, Changes in the
    synthesis of specific proteins following axotomy:
    Detection with two-dimensional gel electrophoresis, J.
    Neurobiol., 9:353-366.
Heacock, A.M., and Agranoff, B.W., 1982, Protein synthesis and
    transport in the regenerating goldfish visual system,
    Neurochem. Res., 7:771-788.
Hendry, I.A., Stockel, K., Thoenen, H., and Iversen, L.L., 1974,
    Retrograde axonal transport of nerve growth factor,
    Brain Res., 68:103-121.
Hirata, F., and Axelrod, J., 1980, Phospholipid methylation and
    biological signal transmission, Science, 209:1082-1090.
Hoffman, S., Sorkin, B.C., White, P.C., Brackenbury, R., Mailham-
    mer, R., Rutishauser, U., Cunningham, B.A., and Edelman,
    G.M., 1982, Chemical characterizations of a neural cell

adhesion molecule purified from embryonic brain membranes, J. Biol. Chem., 257:7720-7729.

Katz, F., Ellis, L., and Pfenninger, K.H., 1984, Nerve growth cones isolated from fetal rat brain: IV. Calcium-dependent protein phosphorylation, in submission.

Laskin, D.L., Laskin, J.D., Weinstein, I.B., and Carchman, R.A., 1981, Induction of chemotaxis in mouse peritoneal macrophages by phorbol ester tumor promotors, Cancer Res., 41:1923-1928.

Letourneau, P.C., 1978, Chemotactic response of nerve fiber elongation to nerve growth factor, Develop. Biol., 66: 183-196.

Llinas, R., and Sugimori, M., 1979, Calcium conductances in Purkinje cell dendrites: Their role in development and integration, in: "Development and Chemical Specificity of Neurons," M. Cuenod, G.W. Kreutzberg and F.E. Bloom, eds., Progr. Brain Res., 51:323-334, Elsevier/North Holland Biomedical Press, Amsterdam.

McGivney, A., Crews, F.T., Hirata, F., Axelrod, J., and Siraganian, R.P., 1981, Rat basophilic leukemia cell lines defective in phospholipid methyltransferase enzymes, Ca^{2+} influx, and histamine release: Reconstitution by hybridization, Proc. Natl. Acad. Sci. USA, 78:6176-6180.

Niedel, J.E., Kuhn, L.J., and Vandenbark, G.R., 1983, Phorbol diester receptor copurifies with protein kinase C, Proc. Natl. Acad. Sci. USA, 80:36-40.

Pfenninger, K.H., and Bunge, R.P., 1974, Freeze-fracturing of nerve growth cones and young fibers. A study of developing plasma membrane, J. Cell Biol., 63:180-196.

Pfenninger, K.H., Ellis, L., Johnson, M.P., Friedman, L.B., and Somlo, S., 1983, Nerve growth cones isolated from fetal rat brain: Subcellular fractionation and characterization, Cell, 35:573-584.

Pfenninger, K.H., and Johnson, M.P., 1981, Nerve growth factor stimulates phospholipid methylation in growing neurites, Proc. Natl. Acad. Sci. USA, 78:7797-7800.

Pfenninger, K.H., and Johnson, M.P., 1983, Membrane biogenesis in the sprouting neuron: I. Selective transfer of newly synthesized phospholipid into the growing neurite, J. Cell Biol., 97:1038-1042.

Pfenninger, K.H., and Maylié-Pfenninger, M.-F., 1981a, Lectin labeling of sprouting neurons: I. Regional distribution of surface glycoconjugates, J. Cell Biol., 89, 536-546.

Pfenninger, K.H., and Maylié-Pfenninger, M.-F., 1981b, Lectin

13

labeling of sprouting neurons: II. Relative movement and appearance of glycoconjugates during plasmalemmal expansion, J. Cell Biol., 89:547-559.

Poole, A.R., Howell, J.I., and Lucy, J.A., 1970, Lysolecithin and cell fusion, Nature, 227:810-814.

Ramón y Cajal, S., 1937, Recollections of my life, translated by E.H. Craigie, with the assistance of J. Cano, in: "Memoirs of the American Philosophical Society," vol. VIII, part 1, pp.368-869, The American Philosophical Society, Philadelphia.

Seeley, P.J., Rukenstein, A., Connolly, J.L., and Greene, L.A., 1984, Differential inhibition of nerve growth factor and epidermal growth factor effects on the PC12 pheochromocytoma line, J. Cell Biol., 98:417-426.

Simkowitz, P., and Pfenninger, K.H., 1983, Rapidly transported proteins of nerve growth cones and synaptic endings are markedly different, Soc. Neurosci. Abst., 9:1178.

Skene, J.H.P., and Willard, M., 1981, Axonally transported proteins associated with growth in rabbit central and peripheral nervous systems, J. Cell Biol., 89:96-103.

Small, R.K. and Pfenninger, K.H., 1984, Components of the plasma membrane of growing axons: I. Spatial distribution of intramembrane particles, J. Cell Biol., 98:in press.

Snyderman, R., and Goetzl, E.J., 1981, Molecular and cellular mechanisms of leukocyte chemotaxis, Science, 213:830-837.

Spitzer, N.C., 1979, Ion channels in development, Ann. Rev. Neuroscience, 2:363-397.

Theiler, R.F., and McClure,.W.D., Rapid axoplasmic transport of proteins in regenerating sensory nerve fibers, J. Neurochem., 31:433-447.

Wallis, I., Ellis, L., and Pfenninger, K.H., 1983, A neuronal polypeptide expressed during neurite formation, J. Cell Biol., 97:242a.

Weinstein, I.B., 1981, Current concepts and controversies in chemical carcinogenesis, J. Supramol. Struct. and Cell Biochem., 17, 99-120.

REGULATION OF SYNAPSE FORMATION BY CYCLIC AMP

M. Nirenberg, K. Krueger, A. Rotter,
S. Wilson, and H. Higashida

Laboratory of Biochemical Genetics, National
Heart, Lung and Blood Institute, National
Institutes of Health, Bethesda, MD 20205

One neuroblastoma and four hybrid cell lines were
obtained which formed numerous synapses with cultured
striated muscle cells. Many other neuroblastoma or
somatic cell hybrids were found which synthesized
acetylcholine, but which lacked one or more properties
needed to form functional synapses. The following kinds
of clonally inherited synaptic defects were identified
in these cell lines: alterations in voltage-sensitive
Ca^{2+} channel activity, large dense-core vesicles,
depolarization-dependent acetylcholine secretion, and
the lack of functional protein synthesized and released
into the medium by the neural cells (which normally
increases aggregation of nicotinic acetylcholine
receptors on myotube plasma membranes).

The number of synapses formed appeared to be a
function of cellular cAMP levels, since prolonged
elevation of cAMP resulted in an increase in the
abundance of synapses in cultures of those cell lines
capable of synaptogenesis. Moreover,in these same cell
lines, prolonged activation of adenylate cyclase with
10 uM PGE_1 (mediated by PGE_1 receptors), or treatment
of cells with 1 mM dibutyryl-cAMP, (which inhibits
cyclic nucleotide phosphodiesterase activity), resulted
in 10-100 fold increases in voltage-sensitive Ca^{2+}
channel activity, 15-50 fold increases in spontaneous
secretion of acetylcholine at synapses, and 5-10 fold
increases in the number of detectable synapses.
Elevation of cellular cAMP levels also resulted in

increases in the number of large dense-cored vesicles, small clear vesicles, and the amount of intracellular acetylcholine. Half-maximal effects on stimulus-secretion coupling, acetylcholine secretion, and synapse abundance were observed when cellular cAMP levels were elevated for 1-2 days, with maximal effects by 5 days. In contrast, elevation of cellular cAMP levels for 20 minutes had no effect on voltage-sensitive Ca^{2+} channel activity, acetylcholine secretion, or numbers of synapses.

Uptake of radiolabelled Ca ($^{45}Ca^{2+}$) by a hybrid cell line (NBr10-A), which is mediated by Ca channels, was inhibited half-maximally by 3 nM nitrendipine. A single class of specific binding sites for [^{3}H]nitrendipine was found in membrane preparations from NBr10-A cells which had been treated with PGE_1 and theophylline, with a dissociation constant of approximately 2×10^{-10}, as estimated by Scatchard analysis. The number of specific sites for nitrendipine was estimated to be 16,000 per cell on the average. In contrast, few if any specific binding sites for [^{3}H]nitrendipine were detected in membranes from untreated, control NBr10-A cells. A species of protein which binds nitrendipine has been reported to be part of the voltage-sensitive Ca^{2+} channel complex. Our results suggest that cAMP regulates the number of voltage-sensitive Ca^{2+} channels, rather than the activity of these channels.

Cyclic AMP-dependent changes in some species of glycoproteins from neuroblastoma-glioma hybrids (NG108-15) were also detected. Elevation of cellular cAMP levels resulted in the appearance or disappearance, as well as changes in abundance or isoelectric points of several different glycoproteins.

In summary, these results indicate that cAMP regulates synapses in cultured neuroblastoma cells and somatic cell hybrids by modulating the expression of a voltage-sensitive Ca^{2+} channel protein which is required for stimulus-secretion coupling and neurotransmitter release. These results also suggest that cAMP regulates both the expression of certain genes and post-translational modification of some species of glycoproteins.

16

MOLECULAR GENETICS OF <u>TORPEDO</u> <u>MARMORATA</u> ACETYLCHOLINE RECEPTOR

A. Devillers-Thiery, J. Giraudat, M. Bentaboulet,
A. Klarsfeld and J.P. Changeux

Institut Pasteur, Laboratoire de Neurobiologie
Moleculaire, 75015 Paris, France

INTRODUCTION

The acetylcholine receptor (AcChoR) is one of the best known
membrane-bound allosteric proteins (reviewed by Changeux, 1981;
Changeux <u>et al.</u>, 1984). Integrated in the post-synaptic membrane
of the cholinergic synapse (neuromuscular junction, electromotor
synapse), it regulates the opening of a cation-selective ionic
channel upon binding of the neurotransmitter (acetylcholine). In
the course of the past fifteen years, the protein assembly which
carries both the acetylcholine binding sites <u>and</u> the ion channel
has been isolated and purified from fish (<u>Electrophorus</u>, <u>Torpedo</u>)
electric organ and vertebrate muscle.

It is a pentameric protein with a molecular mass of about 270
000 daltons composed of four different transmembrane and
glycosylated polypeptide chains with a stoichiometry of 2α, 1β,
1γ, 1δ (reviewed by Karlin, 1980; Changeux <u>et al.</u>, 1984).There
are two acetylcholine binding sites per AcChoR molecule, carried
at least in part by the α-subunits. The receptor protein also
possesses binding site(s) for allosteric ligands, such as the non-
competitive blockers histrionicotoxin, phencyclidine and
chlorpromazine which block the permeability response at a locus
distinct from the acetylcholine binding site. A single high-
affinity site for these non-competitive blockers is present per
oligomer. Covalent labeling experiments suggest that this site is
located in the central pit common to all five subunits (Oswald and
Changeux 1981; Heidmann <u>et al</u>, 1983a, b). Rapid kinetic
measurements of ion fluxes and of ligand binding indicate that
channel opening and closing are regulated by allosteric
transitions between at least four distinct conformational states

of the AcChoR protein (Heidmann et al., 1983b;   Changeux et al., 1984).

At the adult mammalian neuromuscular junction and the electromotor synapse of electric fish, the AcChoR molecules are clustered under the nerve terminal. These densely packed and stable aggregates appear during embryogenesis from a population of more labile and diffusely distributed AcChoR molecules from the surface of the non-innervated myotube or electrocyte. Mammalian muscle embryonic or extrasynaptic AcChoR have also been shown to differ from adult subsynaptic AcChoR by their metabolic turnover, the mean open time of their ionic channel and their immunological reactivity (reviewed in Fambrough, 1979). Several mechanisms have been postulated to account for these differences: expression of different sets of genes at different developmental stages, post-translational modifications of the AcChoR molecule and/or interaction of the AcChoR with extrinsic proteins (such as the ν-1 protein, Saitoh and Changeux, 1980; Gysin et al., 1981; Nghiem et al., 1983).

We have used recombinant DNA techniques in order to study a) the primary and tertiary structure of the AcChoR molecule and b) the regulation of the AcChoR synthesis.

IDENTIFICATION OF cDNA CLONES CODING FOR THE ACETYLCHOLINE BINDING SUBUNIT OF TORPEDO MARMORATA ACETYCHOLINE RECEPTOR

Torpedo electric organ contains 100-fold more AcChoR than mammalian muscle. PolyA$^+$ RNA was purified from adult Torpedo marmorata electric organ. The mRNA species coding for each AcChoR subunit represent 0.2-0.5 % of this preparation, as shown by in vitro translation and specific immunoprecipitation (Mendez et al., 1980; Anderson and Blobel, 1981; Giraudat et al., 1982). Torpedo electric organ mRNA was used to construct a cDNA library in pBR 322 plasmid. The 700 recombinant clones were screened by a two-step procedure: 1) a differential screen; 2) in vitro translation of selected mRNAs. In the first step, recombinant clones were screened for their ability to hybridize to (^{32}P) cDNA complementary to Torpedo electric organ mRNA but not to (^{32}P) cDNA complementary to Torpedo liver and/or Torpedo spleen mRNA. These last two organs contain at least 1000 times less AcChoR than the electric organ. as shown by α-toxin binding site titration.

The forty-six clones that gave a positive signal with electric organ probe and no signal with either liver or spleen probes were further screened by the positive mRNA selection procedure (Ricciardi et al., 1979; Ploegh et al., 1980). In this second step, purified DNA from individual positive clones of the first screen was immobilized on nitrocellulose filters and then hybridized to electric organ mRNA. After dehybridization of each set of filters, the selected mRNAs were translated _in vitro_ and newly synthesized _Torpedo_ AcChoR subunits detected by immunoprecipitation. One such clone yielded a 38K band with both a polyclonal and a monoclonal antibody specific for the denatured α-subunit (Giraudat _et al._, 1982; Tzartos and Lindstrom, 1980). Furthermore, the 38K band disappeared when competing pure α-subunit was added in the immunoprecipitation experiment.This clone, specific for _Torpedo_ AcChoR α-subunit, was designated pα-1. The cDNA insert excised by Pst 1 restriction enzyme from the pα-1 clone was found to contain about 750 bp.

The size of the mRNA species encoding the α-subunit was determined by the "Northern" technique. PolyA$^+$ RNAs from _Torpedo_ electric organ and spleen were size- fractionated by gel electrophoresis, transferred to nitrocellulose filters and hybridized to ^{32}P-labeled pα-1 DNA. One band, about 2000 nucleotides long, was revealed using electric organ mRNA but none with spleen mRNA, thus confirming again the tissue specificity of the AcChoR protein. This also demonstrates that pα-1 DNA does not include the entire mRNA sequence. To obtain longer clones a new cDNA library was constructed and screened with pα-1 DNA ^{32}P-labeled by nick translation. A clone with an insert 1350 bp long was selected and designated pα-2.

NUCLEOTIDE SEQUENCE OF pα CLONES, AND DEDUCED AMINO ACID SEQUENCE OF THE α-POLYPEPTIDE. COMPARISON WITH THE AMINO ACID SEQUENCE OF THE OTHER SUBUNITS

pα-1 and pα-2 cDNAs overlap by 270 bp as shown by their restriction maps (Devillers-Thiery _et al._, 1983). Placed end to end the two cDNA clones represent a unique sequence of 1828 bp, close to the estimated length of the mRNA (2000 NT) encoding the α-subunit. This unique nucleotide sequence (determined according to Maxam and Gilbert, 1980) contains only one open reading frame (Figure 1). The coding sequence, 1383 bp long, is flanked by 76 nucleotides at the 5' end and 369 nucleotides of the 3' untranslated region. It is still incomplete since no polyadenylation site was detected. The corresponding amino acid sequence, 461 residues long, was deduced. The already known N-terminal amino acids of _Torpedo marmorata_ mature α-polypeptide (Devillers-Thiery _et al._, 1979) are indeed present, but are preceded by a short segment of 24 amino acids . This segment contains a large number of hydrophobic residues (58 % of the residues, of which seven leucines), and terminates with a glycine, which has a short side chain. This peptide is characteristic of a transient signal peptide. It is cleaved off from the precursor polypeptide during synthesis and insertion of the protein into the membrane (Anderson and Blobel, 1981) thus yielding a mature α-subunit with 437 amino acids and a molecular weight of 47949 daltons. Except for two positions, this sequence agrees entirely with the partial cDNA sequence of the same species published by Sumikawa et al. (1982).

5' ---AAGAUCUAACAAG

```
                                        -20                          -10                     -1
                        Met Ile Leu Cys Ser Tyr Trp His Val Gly Leu Val Leu Leu Leu Phe Ser Cys Cys Gly Leu Val Leu Gly
UUCCUCGUUUAGUUAUUAGAGUGGCAGAUUUGCUUGAAAAGCCAAUUAUUGAAAGCUGAAGA AUG AUU CUG UGC AGU UAU UGG CAU GUC GGG UUG GUG CUA CUG UUA UUU UCG UGU UGU GGU CUG GUA CUA GGU
      -120              -90            -60                       -30                         -1

  1                   10              Leu Leu Glu Asn Tyr Asn Lys Val Ile Arg 20              Pro Val Glu His His Thr His Phe Val Asp 30 Ile Thr Val Gly Leu Gln Leu Ile Gln 40
Ser Glu His Glu Thr Arg Leu Val Ala Asn Leu Leu Glu Asn Tyr Asn Lys Val Ile Arg Pro Val Glu His His Thr His Phe Val Asp Ile Thr Val Gly Leu Gln Leu Ile Gln Leu
UCU GAA CAU GAA ACA CGU UUG GUU GCU AAU UUA UUA GAA AAU UAU AAC AAG GUG AUU CGU CCA GUG GAG CAU CAC ACC CAC UUU GUA GAU AUU ACA GUG GGG CUA CAG CUG AUA CAA CUC
                 30                              60                              90                            120

Ile Asn Val Asp Glu Val Asn Gln Ile 50 Glu Thr Asn Val Arg Leu Arg Gln Gln Trp 60 Ile Asp Val Arg Leu Arg Trp Asn Pro Ala 70 Asp Tyr Gly Gly Ile Lys Lys Ile Arg Leu
AUC AAU GUG GAU GAA GUA AAU CAA AUU GUG GAA ACA AAU GUG CGC CUA AGG CAG CAA UGG AUU GAU GUG AGG CUU CGC UGG AAU CCA GCC GAU UAU GGU GGA AUU AAA AAG AUC AGA CUG
                 150                             180                             210                           240

Pro Ser Asp Asp Val Trp Leu Pro Asp 90 Leu Val Leu Tyr Asn Asn Ala Asp Gly Asp 100 Phe Ala Ile Val His Met Thr Lys Leu Leu 110 Leu Asp Tyr Thr Gly Lys Ile Met Trp Thr 120 Pro
CCU UCU GAU GAU GUU UGG CUG CCA GAU UUA GUU CUG UAC AAC AAU GCU GAU GGU GAU UUU GCC AUU GUU CAC AUG ACC AAA CUG CUU UGG GAU UAU ACG GGA AAA AUA AUG UGG ACA CCU
                 270                             300                             330                           360

Pro Ala Ile Phe Lys Ser Tyr Cys Glu 130 Ile Ile Val Thr His Phe Pro Phe Asp Gln 140 Gln Asn Cys Thr Met Lys Leu Gly Ile Trp 150 Thr Tyr Asp Gly Thr Lys Val Ser Ile Ser 160 Pro
CCA GCA AUC UUC AAA AGC UAU UGU GAA AUU AUU GUA ACA CAU UUC CCA UUU GAU CAA CAA AAU UGC ACU AUG AAG UUG GGA AUC UGG ACG UAU GAU GGG ACA AAA GUU UCC AUA UCU CCG
                 390                             420                             450                           480

Glu Ser Asp Arg Pro Asp Leu Ser Thr 170 Phe Met Glu Ser Gly Glu Trp Val Met Lys 180 Asp Tyr Arg Gly Trp Lys His Trp Val Tyr 190 Tyr Thr Cys Cys Pro Asp Thr Pro Tyr Leu 200 Asp
GAA AGU GAC CGU CCG GAU CUG AGU ACA UUU AUG GAA AGU GGA GAG UGG GUA AUG AAA GAU UAU CGU GGA UGG AAG CAC UGG GUG UAU UAU ACC UGU UGU CCU GAC ACG CCU UAC CUG GAU
                 510                             540                             570                           600

Ile Thr Tyr His Phe Ile Met Gln Arg 210 Ile Pro Leu Tyr Phe Val Val Asn Val Ile 220 Pro Cys Leu Leu Phe Ser Phe Leu Thr Val 430 Leu Val Phe Tyr Leu Pro Thr Asp Ser Gly
AUC ACC UAC CAU UUU AUC AUG CAG CGU AUU CCU CUU UAU UUU GUG GUG AAC GUC AUC AUU CCU UGU CUG CUU UUU UCA UUU UUA ACU GUA UUA GUA UUU UAC UUA CCA ACU GAU UCA GGU
                 630                             660                             690                           720

Glu Lys Met Thr Leu Ser Ile Ser Val 250 Leu Leu Ser Leu Thr Val Phe Leu Leu Val 260 Ile Val Glu Leu Ile Pro Ser Thr Ser Ser 270 Ala Val Pro Leu Ile Gly Lys Tyr Met Leu 280 Phe
GAG AAG AUG ACU UUG AGU AUU UCC GUU UUG CUG UCU CUG ACU GUG UUC CUU CUG GUU AUU GUU GAG CUG AUC CCC UCA ACU UCC AGC GUG GUG CCU UUG AUU GGC AAA UAC AUG CUU UUU
                 750                             780                             810                           840

Thr Met Ile Phe Val Ile Ser Ser Ile 290 Ile Val Thr Val Val Val Ile Asn Thr His His 300 Arg Ser Pro Ser Thr His Thr Met Pro Gln 310 Trp Val Arg Lys Ile Phe Ile Asn Thr Ile 320
ACA AUG AUU UUU GUC AUC AGU UCA AUC AUC GUU ACU GUU GUU GUA AUU AAU ACU CAC CGC UCU CCA AGU ACA CAU ACA AUG CCA CAA UGG GUA CGA AAG AUC UUU AUU AAU ACU AUA
                 870                             900                             930                           960

Pro Asn Leu Met Phe Phe Ser Thr Met 330 Lys Arg Ala Ser Lys Glu Lys Gln Glu Asn 340 Lys Ile Phe Ala Asp Asp Ile Asp Ile Ser 350 Asp Ile Ser Gly Lys Gln Val Thr Gly Glu 360 Val
CCC AAU CUU AUG UUU UUC UCA ACA AUG AAA CGA GCU UCU AAG GAA AAG CAA GAA AAU AAG AUA UUU GCU GAU GAC AUU GAU AUC UCU GAC AUU UCU GGA AAG CAA GUG ACA GGA GAA GUA
                 990                             1020                            1050                          1080

Ile Phe Gln Thr Pro Leu Ile Lys Asn 370 Pro Asp Val Lys Ser Ala Ile Glu Gly Val 380 Lys Tyr Ile Ala Glu His Met Lys Ser Asp 390 Glu Glu Ser Ser Asn Ala Ala Glu Glu Trp 400 Lys
AUU UUU CAA ACA CCU CUC AUU AAA AAU CCA GAU GUC AAA AGU GCU AUU GAG GGA GUC AAA UAU AUU GCA GAG CAC AUG AAG UCU GAU GAG GAA UCA AGC AAU GCU GCA GAG GAA UGG AAA
                 1110                            1140                            1170                          1200

Tyr Val Ala Met Val Ile Asp His Ile 410 Leu Leu Cys Val Phe Met Leu Ile Cys Ile 420 Ile Gly Thr Val Cys Val Phe Ala Gly Arg 430 Leu Ile Glu Leu Ser Gln Glu Gly
UAU GUU GCA AUG GUG AUU GAU CAC AUU CUG CUG UGU GUC UUC AUG CUG AUU UGU AUA AUU GGU ACA GUU UGC GUG UUU GCU GGC CGU CUC AUU GAA CUC AGU CAA GAG GGC UAA AUCUUCA
                 1230                            1260                            1290                          1320

UUGUGAGCAAAAAAGGCAAUACUGGAAUAAGGGAUGGAUAGCACUCCACAGAAAAGAUGUGUGGGUUUAGUGUUGCAAUUGUAGUCUGUUUUAUGAGAUAUAUAGUUUGCUUUGUUUUUACAAUGAAAUGUACUUAAGGUAUUUGAAUAUGUAAAAAAA
                 1350                     1380                  1410                   1440                  1470

GUAAUGAAAUAACAAUGAGUGAAAAAUGUUAUUAUGCAAGUACCUGAAACAUGUAAUCAGUGGAACAACCUUUUUAAUACAUUACACAAAAGUAAGCAAAAAAUAAGUUUAAAGAAUUUAUGAGGGUAGUCAUUUGAAUUGUAACAGAGAAAUGAAAAUUA
                 1500                 1530                 1560                  1590                  1620

UUAGAAAUAUAAACAGUAUAUAUUUAAGUUAAACAAAGUUAA--- 3'
                 1650           1680
```

Figure 1. Primary structure of the T. marmorata mRNA coding for the α-subunit of the acetylcholine receptor. The mRNA nucleotide sequence was deduced from those of the pα-1 and pα-2 cDNA inserts. Nucleotides are numbered in the 5' to 3' direction starting with the first residue of the codon corresponding to the NH₂-terminal residue of the mature α-subunit. Nucleotides of the 5' untranslated region are indicated by negative numbers. The predicted amino acid sequence of the α-chain precursor is shown above the nucleotide sequence. Amino acid residues are numbered starting with the NH₂-terminal serine of the mature α-subunit. Residues of the transient signal peptide are indicated by negative numbers(From Devillers-Thiery et al.,1983).

There are however eleven nucleotide differences between the
α-subunit cDNA sequences from T. marmorata and T. californica
(Noda et al., 1982). Five of them are silent, six are not and
could very well be species-specific (98,6% homology at the DNA
level between the two species). Since then, cDNA clones encoding
the muscle AcChoR α-subunit from mouse (Merlie et al., 1983) and
calf (Noda et al., 1983a) have been isolated, as well as genomic
clones coding for the human AcChoR α-subunit (Noda et al., 1983a).
The deduced sequences of the α-subunit of Torpedo, calf and human
AcChoR (no sequence is available for the mouse) all contain 437
amino acids and exhibit an extremely high degree of homology
ranging from 80% to 97%. Analysis of the N-terminal sequences
had already revealed striking homology between the four different
subunits of Torpedo californica AcChoR (Raftery et al,1980). It
became apparent that this homology exists all along the
polypeptide chains when the complete amino acid sequences (deduced
from the nucleotide sequences of cDNA clones) became available for
Torpedo californica AcChoR subunits (Ballivet et al., 1983;
Claudio et al., 1983; Noda et al., 1983b, c). This strongly
suggests that the genes encoding them evolved from a common
ancestor by successive gene duplication. (see Noda et al., 1983c).

A MODEL FOR THE TRANSMEMBRANE ORGANIZATION OF THE AcChoR PROTEIN

Straightforward inspection of the amino acid sequence of the
mature α-subunit (Figure 1) discloses four major domains
(Devillers-Thiery et al., 1983). Two of them are rich in
hydrophilic residues (residues 1-209, and 299-408) whereas the
two others have a more hydrophobic character (residues 210-298
and 409-437). These last two more hydrophobic domains can be
further subdivided into four segments containing 20 uncharged
amino acids flanked by charged residues. (I: residues 210-237,
II: residues 243-261, III: residues 277-298, IV: residues 409-
428). As in other transmembrane proteins (Engelman et al., 1980;
Von Heijne, 1981), these are probably lipid spanning α-helices.
Links between segments I, II, III are very short forming a tight
transmembrane core protein. Orientation of the protein in the
membrane was deduced from the results of proteolysis experiments.
Wennogle and Changeux (1980) showed that purified AcChoR-
enriched membrane fragments form sealed vesicles which are
extremely resistant to proteolysis. Degradation of the AcChoR
molecules, which are all oriented right side out, can only occur
after opening of the vesicles. Trypsin then converts the α-
subunit into a 38K fragment, a 35K fragment and finally a stable
32K fragment. The corresponding cleavage sites are thus to be
exposed to the cytoplasmic face of the membrane. Since the N-
terminal amino acid sequence of the 32K fragment is identical to
that of the complete α-chain, the cleavage sites are located near
the C-terminus of the polypeptide. Fourteen lysine and arginine

residues are indeed present in the second hydrophilic domain (residues 299-408) which is thus exposed towards the cytoplasmic face of the membrane. In our model (Figure 2), the N-terminal domain (residues 1-209) representing about 50% of the polypeptide chain would be exposed to the synaptic cleft. Four hydrophobic α-helices (about 25% of the protein mass) would be embedded in the lipid bilayer and the hydrophilic domain, residues 299-408 (25% of the protein mass) would be located on the cytoplasmic side of the membrane. These values are compatible with the values obtained by electron microscopy (Cartaud et al., 1978) or X-ray diffraction (Kistler et al., 1982) with AcChoR, which indicate a total transmembrane height of 11nm with 5.5nm facing the synaptic cleft and a 1.5nm domain extending at the cytoplasmic face.In our model, the C-terminal of the α-polypeptide would be located on the extracellular side of the membrane, which is quite unusual.

The degree of sequence homology between the AcChoR subunits is such that the same structural model can be applied to the β,γ and δ chains. The observed molecular weight differences of the subunits would mainly be accounted for by different sizes of the cytoplasmic domains.

The first hydrophilic domain of the mature α-subunit (residues 1-209) is the only domain facing the synaptic cleft. It must carry the acetylcholine binding site. Indeed, four cysteine residues (residues 128, 142, 192 and 193) are present, two of which could participate in the disulfide bridge located in the vicinity of the acetylcholine binding site (Karlin, 1980). Out of these four residues, cysteine residues number 128 and 142 are also present on the β, γ and δ polypeptides of Torpedo californica (Noda et al., 1983c) whereas cysteine residues number 192 and 193 are only present on the α-polypeptide of the same species. The AcChoR subunits are glycosylated. There are on the N-terminal synaptic domain many possible sites of O-glycosylation (on serine and threonine residues) but there is only one possible site of N-glycosylation (asparagine-141) located next to a cysteine residue.

The available data suggest as the most attractive hypothesis that the ionic channel is located in the center of the AcChoR molecule, and surrounded by all five subunits (Oswald and Changeux, 1981; Heidmann et al, 1983a, b). In this regard, the transmembrane α-helix III from the α-subunit has interesting properties. Inspection of its two-dimensional projection reveals several polar residues amongst the hydrophobic amino acids, located on a line spanning the membrane: tyrosine- 277, threonine - 281, serine - 288, threonine - 292, histidine- 299. A similar arrangement is also possible for the other three subunits of Torpedo AcChoR although it is less evident on the γ and δ

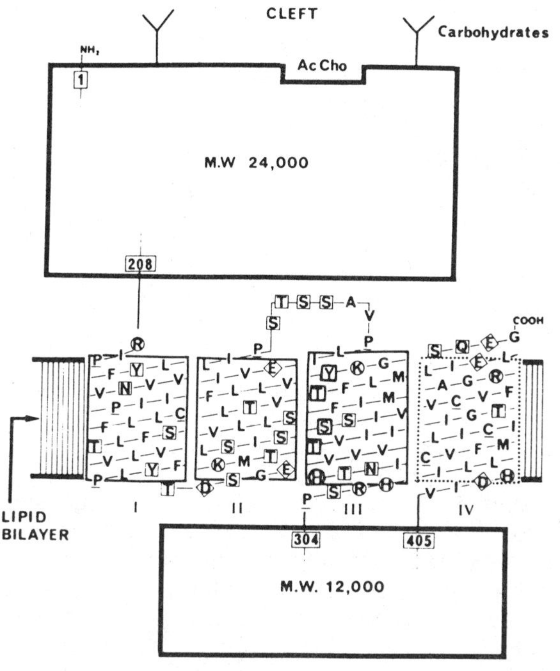

Figure 2. Schematic model of the arrangement of the α-subunit in the postsynaptic membrane. The α-chain is subdivided into three major domains, represented by rectangles having surfaces proportional to their respective molecular weights. The upper rectangle facing the synaptic cleft is the NH$_2$-terminal, mostly hydrophilic, domain (residues 1-208). It carries carbohydrates and at least part of the Ac Cho binding site. The intermediate domain is hydrophobic and composed of four α-helices (I-IV) embedded in the lipid bilayer. The conventional parameters used to construct the helices are helix diameter =9.6 Å, 3.6 residues per turn, and helical pitch =5.4 Å. A planar projection is shown here, and the amino acids are indicated by the one-letter code with uncharged polar, positively charged, and negatively charged residues in squares, circles, and diamonds, respectively, and prolines and cysteines underlined. On helix III, the polar residues organized in the remarkable vertical line (see model) are framed with darker symbols. Since a COOH-terminal post-translational cleavage cannot be excluded (see Devillers-Thiery et al. 1983), helix IV is indicated by a dotted line. The third domain of the α-chain is drawn as a rectangle facing the cytoplasm; this long hydrophilic loop of 101 residues links helix III to IV. A,Ala; C,Cys; D,Asp; E,Glu; F,Phe; G,Gly; H,His; I,Ile; K,Lys; L,Leu; M,Met; N,Asn; P,Pro; Q,Gln; R,Arg; S,Ser; T,Thr; V,Val; W,Trp; Y,Tyr (From Devillers-Thiery et al., 1983).

23

subunits (Noda et al., 1983c). We propose that these five
surfaces of polar residues are exposed towards the inside of the
ionic channel. Transmembrane helices III from each subunit would
thus delineate the channel in its membrane spanning portion. The
internal diameter of such a channel would be close to 6 $\overset{\circ}{A}$ which
is compatible with the estimation of Dwyer et al., 1980. However
cation selectivity cannot be explained solely by the presence of
polar residues in the channel. One has to suppose that exclusion
of anions results from negative charges near the ends of the
channel, probably present on residues of the synaptic and
cytoplasmic domains.

All other authors agree on the assignment of the four
hydrophobic α-helices into the lipid bilayer and on the
localization of the synaptic domain. However, the proposed models
differ as to which peptides form the channel. Noda et al. (1983c)
favor helix I, arguing that the amino acid sequence homology
between α, β, γ and δ polypeptides is higher in that region than in
the other three helices. However one could equally apply the
argument of sequence conservation to regions implicated in
protein-protein interactions which assure the cohesion between
the five subunits. Alternative models have been proposed by
Kossover (1983), Guy (1983) and Finer-Moore and Stroud (1984). In
contrast to our model and the model proposed by Noda et al.
(1983c), they require the presence of charged residues in the
channel which would explain the cation selectivity. They thus
suggest that another transmembrane segment is necessary, the
sequence of which was included in the cytoplasmic domain of our
model. In those three models the C-terminus of the protein faces
the cytoplasmic side of the synaptic membrane.

Despite the different arguments for or against the above
models, there is still no experimental data which demonstrates
unambiguously which one is correct.

A SINGLE GENE ENCODES THE α-SUBUNIT

Gene(s) encoding the α-subunit in Torpedo genome were
detected by the Southern blot technique (Klarsfeld et al, 1984).
Torpedo liver DNA was digested by three restriction enzymes SstI,
Pvu II, and Pst I. DNA fragments were size- fractionated on
agarose gels and transferred to nitrocellulose filters. The DNA
blots were hybridized to four different (^{32}P)-labeled probes
generated by fragmentation of the 1828 bp α-subunit cDNA (pα-1
p α-2) with restriction enzymes. The hybridization patterns
remained identical when blots were washed at high or low
stringency. Two of our probes, one of them entirely in the coding
region, always hybridized to a single genomic fragment. This

result points to the existence of a single gene coding for the
α-subunit. Extensive analysis of the hybridization patterns
showed that no cross-hybridization with the other Torpedo AcChoR
subunits was detected in our conditions. This is expected, since
homology with the other three subunits at the DNA level is below
70 % (our detection limit). The single gene is split by several
introns and has a total length of at least 20 Kb.

Even though no sequence heterogeneity had been observed
at the N-terminal of the α-polypeptide (Devillers-Thiéry et al.,
1979, Raftery et al., 1980) several pieces of data had seemed to
suggest the presence of different α-polypeptides. In the 270 000
dalton oligomer the two sites for acetylcholine and α-toxin show
differences in their affinity for ligands (agonists, competitive
antagonists and some non-competitive blockers). Furthermore,
differences observed between adult and embryonic AcChoR (state
of aggregation, translational motion, mean open time of their
ionic channel and immunological reactivity might reflect sequen
tial expression of different polypeptide chains.A single α - gene
is consistent with the observation that all the individual α -
specific cDNA clones isolated for a given species have the same
sequence (Sumikawa et al. 1982; Devillers-Thiéry et al., 1983;
Noda et al., 1982) Expression of a single gene coding for the α -
subunit at different stages of synaptic development suggests that
the differences observed are "epigenetic" in nature (Changeux and
Danchin, 1976).

In conclusion, cDNA clones coding for the α-subunit of
Torpedo AcChoR have been obtained in our laboratory by
recombinant DNA techniques. The nucleotide sequence of the clones
was determined and the corresponding amino-acid alignments of the
precursor and the mature α-subunit were deduced. From the
primary structure of the α-subunit a model for the transmembrane
organization of the polypeptide in the subsynaptic membrane is
proposed. Since the amino-acid sequences of the other three
subunits show a very high degree of homology with the sequence of
the α-subunit, a similar organization is plausible. In
particular, as suggested by covalent labeling experiments, the
transmembrane portion of the ionic channel is delineated by at
least one α-helix of each subunit. cDNA clones were also used to
probe the gene(s) encoding the α-subunit of Torpedo AcChoR.
A single gene coding for the α-subunit was detected, even though
many pieces of data had suggested the existence of different
polypeptide chains. This result means that differences observed
between embryonic and adult receptors are not due to expression
of different sets of genes during embryogenesis, but rather might
result from post-translational modification of the polypeptides.

REFERENCES

- Anderson, D.J. and Blobel, G. (1981). In vitro synthesis, glycosylation, and membrane insertion of the four subunits of Torpedo acetylcholine receptor.
Proc. Natl. Acad. Sci. USA, 78, 5598-5602.
- Ballivet, M., Patrick, J., Lee, J. and Heinemann, S. (1982). Molecular cloning of cDNA coding for the γ-subunit of Torpedo acetylcholine receptor.
Proc. Natl. Acad. Sci. USA, 79, 4466-4470.
- Cartaud, J., Benedetti, L., Sobel, A. and Changeux, J-P. (1978). A morphological study of the cholinergic receptor protein from Torpedo marmorata in its membrane environment and its detergent extracted purified form.
J. Cell. Sci. 29, 313-337.
- Changeux, J-P. (1981). The acetylcholine receptor: an "allosteric" membrane protein.
The Harvey Lectures 75, 85-254.
- Changeux , J-P., Bon, F., Cartaud J., Devillers-Thiéry, A., Giraudat, J., Heidmann, T., Holton, B., Ngiêm, H.O., Popot, J-L., Van Rapenbusch, R.and Tzartos, S., (1984). Allosteric properties of the acetylcholine receptor protein from Torpedo marmorata .
Cold Spring Harbor Symp. Quant. Biol. 48 (in press).
- Changeux, J-P. and Danchin, A. (1976). Selective stabilization of developing synapses as a mechanism for the specification of neuronal networks.
Nature, 264, 705-712.
- Claudio, T., Ballivet, M., Patrick, J. and Heinemann, S. (1983). Nucleotide and deduced amino acid sequences of Torpedo californica acetylcholine receptor γ-subunit.
Proc. Natl. Acad. Sci. USA, 80, 1111-1115.
- Devillers-Thiéry, A., Changeux, J-P., Paroutaud, P. and Strosberg, A.D.(1979). The amino-terminal sequence of the 40K subunit of the acetylcholine receptor protein from Torpedo marmorata.
FEBS Lett. 104, 99-105
- Devillers-Thiéry, A., Giraudat, J., Bentaboulet, M. and Changeux, J-P.(1983). Complete mRNA coding sequence of the acetylcholine binding α-subunit from Torpedo marmorata acetylcholine receptor. A model for the transmembrane organization of the polypeptide chain.
Proc. Natl. Acad. Sci. U.S.A., 80, 2067-2071
- Dwyer, T.M., Adams, D. and Hille, B. (1980). The permeability of the endplate channel to organic cations in frog muscle.
J. Gen. Physiol. 75, 469-492.
- Engelman, D.M., Henderson, R., Mc Lachlan, A.D. and Wallace, B.A (1980). Path of the polypeptide in bacteriorhodopsin.
Proc. Natl. Acad. Sci. USA, 77, 2023-2027.

- Fambrough, D. (1979). Control of acetylcholine receptors in skeletal muscle.
Physiol. Rev. $\underline{59}$, 165-277.
- Finer-Moore, J. and Stroud, R.M. (1984). Amphiphatic analysis and possible formation of the ion channel in acetylcholine receptor.
Proc. Natl. Acad. Sci. USA, $\underline{81}$, 155-159.
- Giraudat, J., Devillers-Thiéry, A., Auffray, C., Rougeon, F. and Changeux, J-P. Identification of a cDNA clone coding for the acetylcholine binding subunit of _Torpedo marmorata_ acetylcholine receptor.
EMBO J. $\underline{I}$, 713-717.
- Guy, R. (1984). A structural model of the acetylcholine receptor channel based on partition energy and helix packing calculations.
Biophysical J. (in press).
- Gysin, R., Wirth, M. and Flanagan, S. (1981). Structural heterogeneity and sub-cellular distribution of nicotinic synapse associated proteins.
J. Biol. Chem. $\underline{256}$, 11373-11376.
- Heidmann, T., Oswald, R. and Changeux, J-P. (1983a). Le site de liaison de haute affinité de la chlorpromazine est présent à un seul exemplaire par molécule de récepteur cholinergique et est commun aux 4 chaînes polypeptides.
C.R. Acad. Sci. $\underline{295}$, 345-349.
- Heidmann, T., Oswald, R. and Changeux, J-P. (1983b). Multiple sites of action for non-competitive blockers on acetylcholine receptor-rich membrane fragments from _Torpedo marmorata_.
Biochemistry, $\underline{22}$, 3112-3127.
- Karlin, A. (1980). In "Cell Surface Reviews" (Poste, G., Nicolson, G.L. and Cotman, C.W. Eds). Elsevier North Holland Inc. New-York, $\underline{6}$, 191-260.
- Kistler, J., Stroud, R.M., Klymkovsky, M.W., Lalancette R.A. and Fairclough, R.H. (1982). Structure and function of an acetylcholine receptor.
Biophys. J. $\underline{37}$, 371-383.
- Klarsfeld, A., Devillers-Thiéry, A., Giraudat, J. and Changeux, J-P. (1984). A single gene codes for the nicotinic acetylcholine receptor α- subunit in _Torpedo marmorata_: structural and developmental implications.
EMBO J. $\underline{3}$, 35-41
- Kosower, E. (1983). Partial tertiary structure assignments for the β, γ and δ subunits of the acetylcholine receptor on the basis of the hydrophobicity of amino acid sequences and channel location using single group rotation theory.
FEBS Lett. $\underline{155}$, 245-247.
- Maxam, A.M. and Gilbert, W. (1980). Sequencing end-labeled DNA with base specific chemical cleavages.
In Methods in Enzymology $\underline{65}$, 499-560.

- Mendez, B., Valenzuela, P., Martial, J.A and Baxter, J.D (1980). Cell-free synthesis of acetylcholine receptor polypeptides.
Science 209, 695-697.
- Merlie, J-P., Sebbane, R., Gardner, S. and Lindstrom, J. (1983). cDNA clone for the α-subunit of the acetylcholine receptor from the mouse muscle cell line BC3H-1.
Proc. Natl. Acad. Sci. USA, 80, 3845-3849.
- Nghiêm, H.O., Cartaud, J., Dubreuil, C., Buttin, G. and Changeux, J-P. (1983).Production and characterization of a monoclonal antibody directed against the 43,000 MW polypeptide from Torpedo marmorata electric organ.
Proc. Natl. Acad. Sci. U.S.A, 80, 6403-6407.
- Noda, M., Takahashi, H., Tanabe, T., Toyosato, M., Furutani, Y., Hirose, T., Asai, M., Inayama, S., Miyata, T. and Numa, S. (1982). Primary structure of α-subunit precursor of Torpedo californica acetylcholine receptor deduced from cDNA sequence.
Nature, 299, 793-797.
- Noda, M., Furutani, Y., Takahashi, H., Toyosato, M., Tanabe, T., Shimizu, S., Kikyotani, S., Kayano, T., Hirose, T., Inayama, S. and Numa, S. (1983a). Cloning and sequence analysis of calf cDNA and human genomic DNA encoding α-subunit precursor of muscle acetylcholine receptor.
Nature, 305, 818-823
- Noda, M., Takahashi, H., Tanabe, T., Toyosato, M., Kikyotani, S., Hirose, T., Asai, M., Takashima, H., Inayama, S., Miyata, T. and Numa, S. (1983). Primary structures of β and δ subunit precursors of T. californica acetylcholine receptor deduced from cDNA sequences.
Nature, 301, 251-255.
- Noda, M., Takahasi, H., Tanabe, T., Toyosato, M., Kikyotani, S., Furutani, Y., Hirose, T., Takashima, H., Inayama, S., Miyata, T. and Numa, S. (1983). Structural homology of Torpedo californica AchR subunits.
Nature, 302, 528-532.
- Oswald, R. and Changeux, J-P. (1981).Ultraviolet light induced labeling by non-competitive blockers of the acetylcholine receptor from Torpedo marmorata.
Proc. Natl. Acad. Sci., USA, 78, 3925-3929.
- Ploegh, H.L., Ou, H.T., Strominger, J.L (1980). Molecular cloning of a human histocompatibility antigen cDNA fragment.
Proc. Natl. Acad. Sci. USA, 77, 6081-6085.
- Raftery, M.A., Hunkapiller, M., Strader, C. and Hood, L.E. (1980). Acetylcholine receptor: complex of homologous subunits.
Science 208, 1454-1457.
- Ricciardi, R.P., Miller, J.S. and Roberto, B.E. (1979). Purification and mapping of specific mRNAs by hybridization-selection and cell-free translation.
Proc. Natl. Acad. Sci. USA 76, 4927-4931.

- Saitoh, T. and Changeux, J-P. (1980). Phosphorylation in vitro of membrane fragments from Torpedo marmorata electric organ. Effect on membrane solubilization by detergents.
Eur. J. Biochem. 105, 51-62.
- Sumikawa, K., Hougton, M., Smith, J.C., Bell, L., Richards, B.M. and Barnard, E.A. (1982). The molecular cloning and characterization of cDNA coding for the α-subunit of the acetylcholine receptor.
Nucleic Acids Res. 10, 5809-5822.
- Tzartos, S. and Lindstrom, J. (1980). Monoclonal antibodies used to probe acetylcholine receptor structure.
Proc. Natl. Acad. Sci. USA 77, 755-759.
- Von Heijne, G. (1981). Membrane proteins. The amino acid composition of membrane-penetrating segments.
Eur. J. Biochem. 120, 275-278.
- Wennogle, L. and Changeux, J-P. (1982). Transmembrane orientation of proteins present in acetylcholine receptor-rich membranes from Torpedo marmorata studied by selective proteolysis.
Eur. J. Biochem. 106, 381-393.

GENETIC PERSPECTIVES ON BRAIN DEVELOPMENT AND COMPLEXITY

William E. Hahn and Nirupa Chaudhari

University of Colorado
School of Medicine, B-111
Denver, CO 80262

The detection and identification of macromolecular species amidst a vastly complex background is a major problem in neurobiology. Therefore, we wish to begin by making a few introductory remarks pertaining to present estimates of macromolecular complexity of the brain. Genes and their transcripts are the first and second orders respectively of the vast molecular complexity of the brain. The first measurements relevant to estimating the extent to which "single" copy DNA (scDNA or DNA which encodes most of the different proteins) is transcribed in mammalian organs were made about 14 years ago (Hahn, 1970; Hahn & Laird, 1971). These initial measurements, although confirmed by others (Brown & Church, 1971; Grouse, Chilton, & McCarthy, 1972; reviewed by Kaplan & Finch, 1982), were underestimates. But they nonetheless showed that very complex arrays of RNA species are present in eukaryotic cells and organs. We now know that in the mammalian brain (mouse and rat) at least 18-20% of the scDNA is transcribed as nuclear RNA (=~40% of the haploid coding capacity) (Bantle & Hahn, 1976; Chikaraishi, Deeb & Sueoka, 1978). Most of these different transcripts apparently reside in the nuclear RNA of neurons (Ozawa, Kushiya & Takahashi, 1980).

About one-third of the sequence complexity of nuclear RNA is conserved as mRNA (Van Ness, Maxwell & Hahn, 1979; Chikaraishi, 1979). Complexity is defined as the number of nucleotides arranged in sequences which are not repeated in multiple, identical, or very similar copies. Since the haploid scDNA of the mouse genome is 3.2×10^9 nucleotides, the complexity of the mRNA from whole brain is about 2.3×10^8 nucleotides. This is ~55 times more complex than the RNA in E. coli (Hahn, Pettijohn & Van Ness, 1977) and at least 4 times greater than the complexity of mRNA from other complex mammalian organs (Young, Birnie & Paul, 1976; Savage et al., 1978; Van Ness & Hahn, 1980). Linear (nucleotide sequence) complexity comparisons by no means imply proportionality in the actual biological complexity. Orders of complexity arise from the interactive relationships within complex cellular populations, a most notable feature of the brain.

There is difficulty in converting sequence complexity into an estimate of the number of different mRNAs because of variability in their size. Most of the brain mRNA molecules are in the range of 600 - 5000 nucleotides, with a mass average size of 1600 nucleotides. It is possible that most of the different mRNAs are much larger than mass average size. Milner & Sutcliffe (1983) have identified mRNAs, intermediate in overall abundance, which are mostly 3000 -5000 nucleotides in length compared with abundant species which are often in the range of 1500 - 2500 nucleotides. With cloned probes, we have identified individual mRNA species in the range of 800 - 5000 nucleotides (Chaudhari & Hahn, unpublished). If most of the different mRNAs average 4000 nucleotides in length, the number of mRNA species in brain is $2.3 \times 10^8 / 4 \times 10^3$ or ~55,000, as compared to ~137,000 if the mass average size is used.

These considerations tell us that brain mRNA represents a large portion of the DNA which is thought to code for protein (Ohno, 1971; O'Brien, 1973; Rosbash, Campo & Gummerson, 1975). However it should be noted that a substantial number of the nucleotides in mRNA might constitute non-translated regions as shown for several specific messengers (Martin et al., 1981). That more genes are expressed in the brain than in other mammalian organs should come as no surprise. The problems now are to determine which structural genes are activated specifically in the brain, to chart the expression of some of these genes in relation to life cycle and environment, and to determine in which cell types or cellular populations certain genes are active. Fortunately recombinant DNA technology and new techniques for obtaining highly specific antibodies provide probes whereby individual mRNA or protein species may be identified as to location and appearance during development. Hence, despite the complexity of the macromolecular background, molecular genetic dissection of the brain in terms of development and cellular biology seems possible.

BRAIN SPECIFIC MESSENGER RNAs

A fundamental neurobiological problem is the identification of proteins which are unique to the brain. Many such proteins presumably have been selected during the course of evolution to provide structure or function pertinent to the unique biological work performed by the brain. Hence it is important to identify mRNAs which are specific to the brain. This is difficult and

laborious to do with certainty. Thus, as a first approximation we
have assumed, as have others, that mRNA species which are detected
in brain polysomal RNA, but are absent in other complex organs such
as the liver, kidney, spleen, etc., are brain specific. Given this
limited criterion, a large number of brain specific mRNAs have
been identified.

In the rodent brain, two major complex classes of mRNA are
present, namely polyadenylated (poly(A)$^+$) and nonpolyadenylated
(poly(A)$^-$) mRNAs. The criteria for identifying a population of
RNA molecules as messenger are inclusion in polyribosomes, translatability, and high sequence complexity. First, we describe some
of the features of the poly(A)$^+$ mRNAs.

Brain Poly(A)$^+$ mRNAs

Poly(A)$^+$ mRNAs are present roughly in 3 major copy frequency
classes, namely, abundant, intermediate, and rare. Arbitrarily,
abundant species are present $10^3 - 10^4$ copies/cell or greater,
intermediate species are $10^2 - 10^3$ copies/cell, and rare class
species are ~1 - 10 copies/cell averaged over the whole brain.
Abundant species are probably present in all or most cells. Many
of the rare species are probably restricted to particular cell
types or given anatomical regions of the brain where they may be
present in relatively high copy number. Species which are both
rare and restricted would fall far below one copy per cell on a
global basis, and would not be detected by the methods we have
used.

A large number of unidentified poly(A)$^+$ mRNAs in the brain
are encoded by genes containing introns ("split" genes) such that
mRNAs are not co-linear with the respective transcribed DNA
(Maxwell, Maxwell & Hahn, 1980), as has been shown for a number of
specific eukaryotic genes (Breathnach, Mandel & Chambon, 1977;
Jeffreys & Flavell, 1979; Sargent et al., 1979).

Brain poly(A)$^+$ mRNAs constitute slightly over half the
sequence complexity of total mRNA (Van Ness, Maxwell & Hahn, 1979;
Chikaraishi, 1979). It appears that ~60% of the different
sequences are specific to the brain. The latter has been shown by
hybridization measurements with cDNA (DNA copied from RNA by
reverse transcription) representing the so-called complex or rare
class of brain mRNAs. Polysomal RNA from the liver and kidney only
hybridize a portion of this cDNA indicating that a large number of
mRNAs are "specific" to the brain (Table I). Similar measurements
with nuclear RNA from liver and kidney indicate that most of the
brain specific poly(A)$^+$ mRNAs are transcriptionally regulated.
However, the somewhat greater maximal hybridization of the cDNA
with nuclear RNA indicates that some genes specifying brain
restricted poly(A)$^+$ mRNAs are transcribed in liver and kidney

TABLE 1. Brain Specific Poly(A)$^+$ mRNAs and Transcriptional Control

RNA	Rare class cDNA hybridized %
Brain total polysomal	100
Liver polysomal	32
Kidney polysomal	38
Brain Nuclear	100
Liver Nuclear	48
Kidney Nuclear	51

(leaky or constitutive transcription?), but these transcripts are not detectably conserved as mRNA, perhaps owing to posttranscriptional regulatory mechanisms (Davidson & Britten, 1979; Shepard & Nemer, 1980; Xin et al., 1982).

Measurements as summarized in Table 1 pertain to the rare, complex class of mRNA. Brain specific mRNA species of the abundant and intermediate classes have been identified by screening a brain cDNA library. In Figure 1, clones containing cDNA complementary to brain poly(A)$^+$ mRNAs are shown; examples of putatively brain species are boxed. Clone B1-6:9, for example is shown to be brain specific in that it cannot be detected in RNA from liver or kidney at the resolution of northen blot analysis (Figure 1-b). Screening a cDNA library made from rat and mouse brain poly(A)$^+$ mRNAs showed that ~30% of the selected clones representing abundant to intermediate class mRNAs were specific to the brain as determined by northern blots (Milner & Sutcliffe, 1983; Chaudhari & Wilber, unpublished data). Thus, brain specific mRNAs are found in all the major abundance classes.

Single strand probes (produced by recloning or other methods) suitable for exhaustive probe driven hybridization can be used to establish whether mRNAs shown to be restricted by northern blot hybridizations are in fact truly brain specific. Probes of this sort are also useful for in situ hybridization to determine the cellular location of individual mRNAs.

Most of the poly(A)$^+$ mRNA species present in adult brain are present at birth. This is known from experiments showing that the sequence complexity of newborn poly(A)$^+$ mRNA is similar to that of the adult (Chaudhari & Hahn, 1983). We point out the measurements which have been made are sensitive only to within 3000 - 4000 different, average size mRNA species. With cDNA probes, we have found that a substantial number of the poly(A)$^+$ mRNAs are absent in

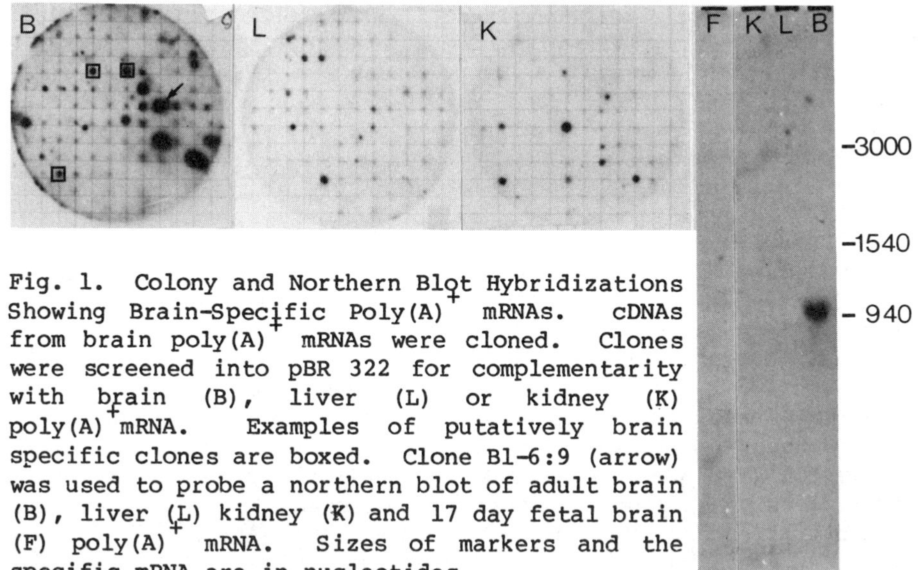

Fig. 1. Colony and Northern Blot Hybridizations Showing Brain-Specific Poly(A)$^+$ mRNAs. cDNAs from brain poly(A)$^+$ mRNAs were cloned. Clones were screened into pBR 322 for complementarity with brain (B), liver (L) or kidney (K) poly(A)$^+$ mRNA. Examples of putatively brain specific clones are boxed. Clone B1-6:9 (arrow) was used to probe a northern blot of adult brain (B), liver (L) kidney (K) and 17 day fetal brain (F) poly(A)$^+$ mRNA. Sizes of markers and the specific mRNA are in nucleotides.

the fetal brain 4 days prior to birth (Chaudhari & Hahn, 1983). By screening a cDNA library representing mRNA from adult brain, we have identified highly abundant mRNAs which are either absent or present in low concentrations in the fetus compared to adult, and vice versa. For example, clone B1-6:9 (arrow) shown in Figure 1, represents an mRNA species which is present in adult brain, but is undetectable by northern blot in fetal brain (17 day). In situ hybridization shows that this cloned cDNA is complementary to mRNA which is abundant in white matter, and thus it may be a species prevalent in glial cells, but absent or present in much reduced amounts in neurons. The appearance of this mRNA after birth, its relative abundance, and its localization in white matter suggests it might be specifying a protein important in myelinization.

Thus, libraries may be screened for mRNAs with a number of different criteria in mind such as appearance during development, specificity to the brain, anatomical distribution, cell type restriction, relative abundance, size, etc. It should therefore be possible to place genetic expression in the brain in context with cellular and developmental biology.

Brain Poly(A)$^-$ mRNAs

In contrast with the poly(A)$^+$ mRNAs, the genes specifying brain poly(A)$^-$ mRNAs are mostly activated during the course of postnatal development. Additionally, nearly all of the different poly(A)$^-$ mRNAs species are specific to the brain. In fact, the presence of a complex population of poly(A)$^-$ mRNA appears to be unique to the brain in that such a population has not been detected

in other complex organs (Bantle , Courchesne & Couch, 1980;
Ouellette et al., 1982; Chaudhari & Hahn, 1983) or cultured
mammalian cells (Maxwell & Hahn, unpublished data).   Poly(A)$^-$
mRNAs which are common among organs are mostly highly abundant
species which collectively constitute only a small fraction of the
total sequence complexity. However, molecular cloning experiments
show that brain specific poly(A)$^-$ mRNA species can be identified
which are intermediate in abundance.

We now wish to summarize some data pertaining to the complexity of the poly(A)$^-$ mRNA population and the postnatal expression
of respective genes. As given in Table 2, the sequence complexity
of adult polysomal RNA is essentially twice that of the poly(A)$^+$
mRNA fraction alone. Thus, poly(A)$^-$ mRNAs similar in complexity
to poly(A)$^+$ mRNAs constitute a separate non-overlapping population
of sequences. Measurements with cDNA probes also lead to the same
conclusion (Van Ness, Maxwell & Hahn, 1979). For technical considerations pertaining to complexity measurements based on nucleic
acid hybridization. See Van Ness & Hahn (1980, 1983).

As given in Table 2, the complexity of newborn total nuclear
RNA is less than that of the adult, but the poly(A)$^+$ fractions
(poly(A)$^+$ hnRNA) are nearly the same. Thus, during postnatal
development, the increase in nuclear RNA complexity (Table 2) is
attributable to the appearance of transcripts which lack poly(A).
Some or most of these postnatally appearing nuclear RNAs are logi-
cally the precursors of poly(A)$^-$ mRNAs.

TABLE 2. Sequence Complexity Of Brain RNAs In Relation
To Development

RNA	Age	Sequence Complexity in Nucleotides
Total polysomal	Newborn	1.2×10^8
Poly(A)$^+$ mRNA	Newborn	1.0×10^8
Total polysomal	Adult (60 days)	2.3×10^8
Poly(A)$^+$ mRNA	Adult	1.2×10^8
Total Nuclear	Newborn	4.2×10^8
Poly(A)$^+$ hnRNA	Newborn	4.0×10^8
Total Nuclear	21 day	5.2×10^8
Total Nuclear	Adult	5.6×10^8
Poly(A)$^+$ hnRNA	Adult	4.0×10^8

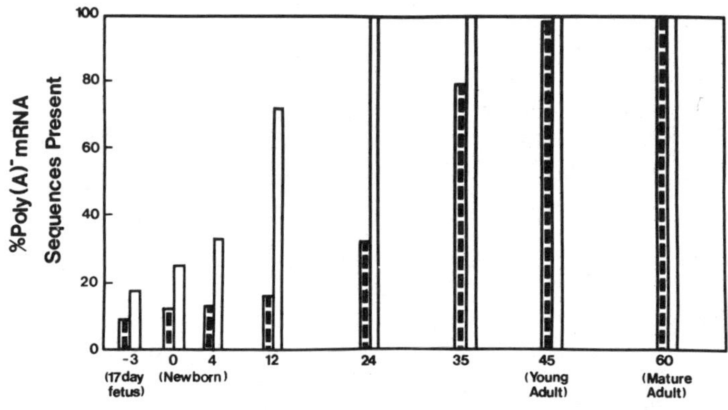

Fig. 2.   Appearance of Brain Poly(A)⁻ mRNA Sequence During Development. Open bars, nuclear RNA; hatched bars, polysomal RNA. Taken from Chaudhari & Hahn, 1983.

With cDNA representing the rare copy, complex class of poly(A)⁻ mRNAs we have probed whole brain polysomal RNAs prepared from mice of various ages. We have found that about 90% of the different sequences appear after birth. The appearance of poly(A)⁻ mRNAs in polysomal RNA is summarized in Figure 2. Notice that the appearance is gradual, extending to young adulthood.

Puzzling, and without any precedent for explanation, is the observation that a large portion of the poly(A)⁻ mRNAs are measurable in the nuclear RNA (i.e., genes are transcribed) well in advance of their appearance in the polysomes (Fig. 2). Nor are they simply present in the cytoplasm as free ribonucleoprotein particles as shown by other experiments (Chaudhari & Hahn, 1983). Notice at 24 days most of the poly(A)⁻ mRNA sequences are present in the nuclear RNA, whereas only ~30% can be detected in polysomes. This implies the presence of a posttranscriptional control which governs the release of these mRNAs to the cytoplasm. Posttranscriptional controls are apparent from studies on developing sea urchins (Wold et al., 1978) and implied from the presence of brain specific mRNA sequences in liver and kidney nuclear RNA (see Table 1). The long delay suggests that developmental events, and perhaps activities triggered by environmental inputs arriving via various sensory modalities, are necessary to modulate posttranscriptional control mechanisms leading to the release of various mRNAs. But interpretation of this observation is purely speculative. The matter remains perplexing and mysterious.

Essentially nothing is known regarding activation, coordinate control, etc., of genes specifying brain specific mRNAs during embryonic and postnatal development. Recently, an 82 nucleotide sequence, repeated more than 10^5 times per genome, has been

reported to be associated with genes specifying brain restricted
poly(A)$^+$ mRNAs in the rat (Sutciffe et al., 1982). This so-called
ID (identifier) sequence might have a general regulatory function
common to many, perhaps all, of the genes specifying brain
specific poly(A)$^+$ mRNAs. It is possible that the same, or an
analogous, sequence is found near or in genes encoding brain
specific poly(A)$^-$ mRNAs.

Protein Complexity In Relation To Polysomal RNA Complexity

Evidence for the existence of most of the different proteins
inferred from measurements of mRNA complexity is, of course, lacking. Although 2-D electrophoretic analysis of the in vitro translation products of brain mRNAs show numerous polypeptides, the
number is vastly less than the estimated number of mRNAs as proteins encoded by rare class species are not resolved (Chikaraishi,
1979; Morrison, Pardue & Griffin, 1981). However, it is possible
to assess the relative translatability of the rare, complex class
poly(A)$^+$ and poly(A)$^-$ mRNAs. We did this by selectively arresting
the translatability of abundant mRNA species by hybridizing them
with cDNA, leaving the rare class species available as template in
an in vitro translation system. On a per mass basis, the rare
class poly(A)$^+$ and poly(A)$^-$ mRNAs were found to be 80 - 100% as
effective as total mRNA in a reticulocyte lysate system (Hahn et
al., 1983). This experiment does not, of course, establish that
these mRNAs are represented by respective proteins in the brain,
it merely establishes potential translatability. In the in vivo
state many of the mRNAs could be arrested from translatability in
some manner such as association with blocking proteins.

In order to establish alignment between brain specific mRNAs
and brain proteins we have begun nucleotide sequencing of selected
clones. The presence of an open reading frame (an extended
sequence uninterrupted by stop codons) implies mRNA function. For
example, we have identified an open reading frame in a cDNA clone
representing a brain specific poly(A)$^-$ mRNA. The nucleotide
sequence of this reading frame can be used definitively to establish the presence of a protein, preferably in situ, corresponding
to this specific mRNA species. This can be accomplished by synthesizing peptides corresponding to different regions of the putative
protein. The various peptides, usually 10 - 20 amino acids in
size, are then conjugated to carrier molecules and used as immunogens to elicit antibodies. The immunoglobulins usually recognize
the peptide as it resides in the intact protein (reviewed by
Arnon, 1980; Lerner, 1982; Shinnick et al., 1983). Antibodies
obtained by this approach have already been applied with remarkable success to demonstrate the existence and location of unknown
proteins specified by relatively abundant rat brain specific mRNAs
(Sutcliffe et al., 1983). If immunoglobulins specific to two or
more well separated peptide-sized sequences within the putative

protein are used, the protein can be identified unambiguously with a specific mRNA. This approach should lead to the detection of proteins specified by mRNAs, which in a global sense are rare copy, but which are relatively abundant in particular groups or networks of cells. Also, this approach may lead to identification of proteins which are relatively abundant but owing to very low turnover, their respective mRNAs remain at low levels.

If proteins can be demonstrated for a few, selected rare class mRNAs, the assumption that protein diversity (at least in terms of amino acid sequence) is essentially equivalent to mRNA code sequence complexity would be justified. Of course, owing to alternate RNA processing and posttranslational modifications, a wider variety of polypeptides may be present than estimated simply from linear sequence complexity. Also, it is possible that certain genes or families of genes are subject to recombinatorial events analogous to those occurring in the immunoglobulin genes (Seidman & Leder, 1978; Weigert et al., 1978). Such added diversity would not be detected by linear complexity measurements.

QUESTIONS, PERSPECTIVES, AND SPECULATIONS

To understand the mammalian brain is clearly the ultimate challenge in scientific inquiry. The study of rudimentary nervous systems and "simple" brains of various species may provide basic clues. Over a wide range of animal species, from molluscs to man, common cellular properties and anatomical structures can be identified in the nervous system. Although common properties imply the existence of unifying biological rules and mechanisms, it is clear that huge differences in behavioral ability exists. Besides gross differences such as size and general architecture, what is the nature and magnitude of the differences which account for the enormous variation in behavioral capacity? Are the seemingly quantal, if not unique, jumps in the evolution of intelligence underpinned by the selective emergence of many new genes and gene families, or are comparatively few genetic changes adequate for the great differences apparent amongst species?

While it is clear that the brain is the most complex organ, data on the comparative molecular biology of this organ are very limited. Data similar to that obtained from rodent brain are not available for other vertebrates, such as fish, amphibians, birds--or man, although preliminary data indicates RNA complexity is less in fish brain than in mouse (B. Kaplan, personal communication) and very high in man (Grouse, Omenn & McCarthy, 1973). In vertebrates, is there simply a similar number of genes encoding similar proteins functioning in similar ways, or has there been selection for many "new" additional genes as new areas or neuronal populations emerged in the brain during the course of evolution? If the latter is the case, and we think it is, many of the brain-

and species-specific proteins would be the most recent products of
evolution in comparison with those specific to other organs such
as liver and kidney where the basic function and complexity
required seemingly was selected early on, being roughly the same
in most vertebrates. Perhaps, most of the significant structural
gene differences between vertebrate species in general, or more
specifically between man and other primates, is attributable to
genes specifying proteins which are unique to the brain (see
Lumsden & Wilson, 1981, for evolutionary considerations).
Clearly, behavioral specialization as well as general behavioral
versatility figured greatly in speciation. How could this not
have been, or continue to be, the case? In particular, genes which
are activated during postnatal development might be the genes
where the greatest differences amongst species are found, particularly in regard to those species in which postnatal development is extensive, elaborate, and highly dependent upon sensory
input and experience.

Although the power and increasing precision of molecular
genetics is impressive, as evidenced by numerous recent successes,
we cannot reasonably expect to understand the subtle cellular
orchestrations and intricacies of the brain by this approach. To
attempt an understanding of the brain in the unitary sense of
complex behavior will obviously require the convergence of several
scientific disciplines and philosophy. But we should expect a lot
from molecular genetics. For one, it can provide probes useful in
delineating cellular relationships. The availability of an
increasing number of probes will greatly accelerate advances in
cytological neuroanatomy. We are on our way, at a rapidly increasing rate, to identifying and characterizing many of the molecular
species and receptors essential for brain function, and we should
be able to place them in functional context with interactive cellular networks of the brain. This alone will be a major task.

REFERENCES

Arnon, R., 1980, Chemically defined antiviral vaccines, Ann.
    Rev. Microbiol., 34:593.
Bantle, J. A., Courchesne, C. L., and Couch, M., 1980, Complexity
    and complexity overlap in mouse liver polyadenylated and
    nonadenylated messenger RNA fractions, Biochem. Biophys. Res.
    Commun., 95:1710.
Bantle, J. A., and Hahn, W. E., 1976, Complexity and characteri-
    zation of polyadenylated RNA in the mouse brain, Cell,
    8:139.
Brown, J. R., and Church, R. B., 1971, RNA transcription from
    nonrepetitive DNA in the mouse, Biochem. Biophys. Res.
    Commun., 42:850.
Chaudhari, N., and Hahn, W. E., 1983, Genetic expression in the
    developing brain, Science, 220:924.

Chikaraishi, D. M., 1979, Complexity of cytoplasmic polyadenylated and nonpolyadenylated rat brain ribonucleic acids, Biochem., 18:3249.

Chikaraishi, D. M., Deeb, S. S., and Sueoka, N., 1978, Sequence complexity of nuclear RNAs in adult rat tissues, Cell 13:111.

Davidson, E. H., and Britten, R. J., 1979, Regulation of gene expression: Possible role of repetitive sequences, Science, 204:1052.

Grouse, L. D., Chilton, M. D., and McCarthy, B. J., 1972, Hybridization of ribonucleic acid with unique sequences of mouse deoxyribonucleic acid, Biochem. 11:798.

Grouse, L. D., Omenn, G. S., and McCarthy, B. J., 1973, Studies by DNA-RNA hybridization of transcriptional diversity in human brain, J. Neurochem., 30:191.

Hahn, W. E., 1970, Transcription of nonrepeated DNA in brain, J. Cell Biol., 10:31.

Hahn, W. E., Chaudhari, N., Beck, L., Wilber, K., and Peffley, D., 1983, Genetic expression and postnatal development of the brain: Some characteristics of nonpolyadenylated mRNAs, Cold Spring Harbor Symposia, Vol. 48, Chap. 24, Cold Spring Harbor, NY.

Hahn, W. E., and Laird, C. D., 1971, Transcription of nonrepeated DNA in mouse brain, Science 173:158.

Hahn, W. E., Pettijohn, D. E., and Van Ness, J., 1977, One strand equivalent of the Escherichia coli genome is transcribed: Complexity and abundance classes of mRNA, Science, 197:582.

Kaplan, B. B., and Finch, C. E., The sequence complexity of brain ribonucleic acids. IN: "Molecular Approaches to Neurobiology," I. R. Brown, ed., Academic Press, N.Y., (1982).

Lerner, R. A., 1982, Tapping the immunological repertoire to produce antibodies of predetermined specificity, Nature, 299:592.

Lumsden, C. J., and Wilson, E. O., 1981, Genes, mind and culture, Harvard University Press, Cambridge.

Martin, S. L., Zimmer, E. A., Davidson, W. S., Wilson, A. C., and Kan, Y. W., The untranslated regions of β-globin mRNA evolve at a functional rate in higher primates, Cell, 25:737.

Maxwell, I. H., Maxwell, F., and Hahn, W. E., 1980, General occurrence and transcription of intervening sequences in mouse genes expressed via polyadenylated mRNA, Nuc. Acids Res., 8:5875.

Milner, R. J., and Sutcliffe, J. G., 1983, Gene expression in rat brain, Nuc. Acids Res., 11:5497.

Morrison, M. R., Pardue, S., & Griffin, W. S. T., 1981, Developmental alterations in the levels of translationally active mRNAs in the postnatal rat cerebellum, J. Biol. Chem., 256(7):3550.

O'Brien, S. J., 1973, On estimating functional gene numbers in eukaryotes, Nature, 242:52.

Ohno, S., 1971, Simplicity of mammalian regulatory systems inferred by single gene determinations of sex phenotypes, Nature, 234:134.

Ouellette, A. J., Ordahl, C. P., Van Ness, J., and Malt, R. A., 1982, Mouse kidney nonpolysomal messenger RNA: Metabolism, coding function and translational activity, Biochem., 21:1169.

Ozawa, H., Kushiya, E., and Takahashi, Y., 1980, Complexity of RNA from the neuronal and glial nuclei, Neurosci. Lett., 18:191.

Rosbash, M., Campo, M.S., and Gummerson, K. S., 1975, Conservation of cytoplasmic poly(A)-containing RNA in mouse and rat, Nature, 258:582.

Savage, M. J., Sala-Trepat, J. M., and Bonner, J., 1978, Measurement of the complexity and diversity of poly (adenylic acid) containing messenger RNA from rat liver, Biochem., 17:462.

Seidman, J. G., and Leder, P., 1978, The arrangement and rearrangement of antibody genes, Nature, 276:790.

Shepherd, G. W., and Nemer, M., 1980, Developmental shifts in frequency distibution of polysomal mRNA and their post-transcriptional regulation in the sea urchin embryo, Proc. Natl. Acad. Sci., USA, 77:4653.

Shinnick, T. M., Sutcliffe, J. G., Green, N., and Lerner, R. A., 1983, Synthetic peptide immunogens as vaccines, Ann. Rev. Microbiol., 37:425.

Sutcliffe, J. G., Milner, R. J., Shinnick, T. M., and Bloom, F. E., 1983, Identifying the protein products of brain-specific genes with antibodies to chemically synthesized peptides, Cell, 33:671.

Sutcliffe, J. G., Milner, R. J., Bloom, F. E., and Lerner, R. A., 1982, Common 82-nucleotide sequence unique to brain RNA, Proc. Natl. Acad. Sci., 79:4942.

Van Ness, J., and Hahn, W. E., 1980, Sequence complexity of cDNA transcribed from a diverse mRNA population, Nuc. Acids Res., 8:4259.

Van Ness, J., and Hahn, W. E., 1983, Physical parameters affecting the rate and completion of RNA driven hybridization of DNA: New measurements relevant to quantitation based on kinetics, Nuc. Acids Res., 10:8061.

Van Ness, J., Maxwell, I. H., and Hahn, W. E., 1979, Complex population of nonpolyadenylated messenger RNA in mouse brain, Cell, 8:1341.

Weigert, M., Gatmaitan, L., Loh, E., Schilling, J., and Hood, L., 1978, Rearrangement of genetic information may produce immunoglobulin diversity, Nature, 276:2785.

Xin, J., Brandhorst, B. P., Britten, R. J., and Davidson, E. H., 1982, Cloned embryo mRNAs not detectably expressed in adult sea urchin coelomocytes, Devel. Biol., 89, 527.

Young, B. D., Birnie, G. D., and Paul, J., 1976, Complexity and specificity of polysomal poly(A)$_4$ mRNA in mouse tissues, Biochem., 15:2823.

GENOMIC AND PHENOTYPIC EXPRESSION OF AUTONOMIC NEURONS

Tong H. Joh

Laboratory of Neurobiology
Cornell University Medical College
1300 York Avenue
New York, NY 10021

INTRODUCTION

As early as 1909 Elliott[1] proposed that an adrenaline–like substance is released from sympathetic nerve terminals to produce a chemical neurotransmission. Since then, various chemical neurotransmitters have been identified. The classical chemical neurotramsnitters, catecholamines (CA), serotonin (5HT), acetylcholine (ACh) and gamma-aminobutyric acid (GABA), are most venerable neurotransmitters.

The phenotype of each neuron can be identified by the specific enzyme(s) it uses to produce its own neurotransmitter, and the specific neurotransmiter high affinity uptake site(s) found at its nerve endings. Neurotransmitters are stored in their specific storage vesicles, and they are released by exocytosis[2]. It is not know whether different neurotransmitters are stored in different storage granules. It is also not known whether the membrane structures of different neurotransmitter storage vesicles are different. It is known, however, that when the neurotransmitter is released by exocytosis, the inner membrane of the vesicle become exposed to the outer membrane of the nerve endings. For instance, dopamine B-hydroxylase (DBH), which catalyzes the conversion of dopamine to norepinephrine in noradrenergic neurons, is attached to or integrated into the inner membrane of the vesicles or chromaffin granules of adrenal medulla[3] (a membrane bound form of DBH). When norepinephrine is released by exocytosis, membrane bound DBH is exposed to the outer membrane of the nerve endings[3]. Production of the norepinephrine synthesizing enzyme, DBH, and the association of DBH with noradrenergic nerve endings, are the structural characteristics of noradrenergic phenotype. It is of interest to investigate whether this kind of structural characteristic of neuronal phenotypes exists in other neurons. We therefore sought to investigate the relationships between genes coding for neurotransmitter enzymes and certain membrane protein(s).

GENES FOR CATECHOLAMINE BIOSYNTHETIC ENZYMES

Over the past twenty years, we have been studying the mechanisms governing the regulation of the enzymes catalyzing the biosynthesis of neurotransmitters. These enzymes are tyrosine hydroxylase (TH), DBH and phenylethanolamine N-methyltransferase (PNMT) for catecholamine (CA) biosynthesis, tryptophan hydroxylase for serotonin synthesis, and choline acetyltransferase (ChAT) for the biosynthesis of acetylcholine (ACh). These enzymes have been purified to homogeneity and antibodies to these enzymes have been raised[4-7]. Using immunochemical techniques, we have studied the regulatory mechanisms of enzyme induction in the central nervous system and/or neuronal tissues and cells[8-13].

The focus of our research for many years has been the regulation of CA biosynthetic enzymes. An important feature of the CA biosynthetic pathway is that all four enzymes, including dopa decarboxylase (DDC), are not expressed together in all cells. While all are expressed in neurons or chromaffin cells synthesizing epinephrine, norepinephrine containing neurons or chromaffin cells do not express PNMT. In cells producing DA as their final neurotransmitter, neither DBH nor PNMT are present. Based on the classical one gene for one enzyme hypothesis, these facts would suggest that separate genes code for each enzyme and that these genes are probably independently regulated. However, our recent findings indicate that genes for TH, DBH and PNMT share the same gene coding sequences[14], suggesting common ancestry and perhaps some coordinated regulatory mechanisms among them. In the present studies, we will demonstrate that not only CA enzymes share similar protein domains in their primary structures, but also the neurotransmitter enzyme and membrane protein of its own synaptosome may share common protein domains.

MOLECULAR WEIGHTS OF CA BIOSYNTHETIC ENZYMES

Molecular weights of TH, DBH and PNMT determined by Sodium dodecylsulfate (SDS) containing polyacrylamide gel electrophoresis (PAGE) are 60, 75 and 31 kilodaltons (kd), respectively[5]. When purified TH was digested by trypsin for 5 min at room temperature, an enzymatically active and stable form of TH was isolated[3,15]. This trypsin digested form of TH (tTH) has a MW of 35 kd[3].

PEPTIDE MAPPING OF TH, DBH AND PNMT

Digestion of CA synthesizing enzymes with CNBr and chymotrypsin produced 55K proteins from both TH and DBH, and amino acid composition of the 55K peptides was very similar[14]. Digestion of these enzymes

44

with _Staphylococcus_ _aureus_ V8 protease produced several identical peptides. Major peptides included a 10K moiety common to DBH and TH, and a 20K peptide common to DBH and PNMT. A somewhat diffuse minor band appeared among all three enzymes and a minor 15K peptide was common to PNMT and DBH. The 10K and 20K peptides were eluted from an identical unstained gel and were analyzed for amino acid content. An amino acid analysis revealed a striking co-relation, indicating that the 10K peptides in TH and DBH as well as the 20K peptides in DBH and PNMT are nearly identical[14].

For an extensive proteolysis, the purified enzyme was digested with trypsin, and the trypsin digests were then analyzed by reversed phase HPLC. Peptide fractions held in common among TH, DBH and PNMT were concentrated and hydrolyzed for amino acid analysis.

The retention times of peptides separated from TH and DBH digests show a striking similarity[14]. Twenty-one of the 32 peaks in TH elute at times comparable to the appearance of peptides in DBH. Amino acid composition of the 21 peptides common to TH and DBH revealed a high degree of homology between these two enzymes.

Similar analyses of peptides and amino acid composition of the peptides revealed that homology between PNMT and TH, and PNMT and DBH is less than that of TH and DBH. However, approximately 25 to 30% homologies exist among these enzymes. The results suggest that TH, DBH and PNMT share common protein domains in their primary structure.

CROSS-REACTIVITY OF ANTIBODIES WITH mRNA TRANSLATION PRODUCTS

Polysomal poly(A)mRNA was purified from bovine adrenal medulla and the mRNA was translated _in vitro_ in a reticulocyte lysate translation system as described previously[8,16]. Total bovine adrenal mRNA was isolated as described by Seeburg et al.[17]. Immunoprecipitation of CA enzymes from _in vitro_ poly(A)mRNA translation products was carried out by the procedures detailed in our previous publications[4,8,14,16].

Antibodies raised against trypsin digested form of TH (tTH) reacted predominantly with a 31 kilodalton protein, identical in size to PNMT[14]. When an antibody raised against the native form of rat adrenal TH was used, coprecipitation of a small amount of a 72 kilodalton protein in addition to the 60 kilodalton TH was observed[14]. This 72 kilodalton protein is identical in size to the protein subunit precipitated with DBH antibody. In addition, bovine PNMT antibody coprecipitated a small amount of a 60 kilodalton protein, identical in size to TH[14]. The results further strengthen our hypothesis that CA synthesizing enzymes share common protein domains in their primary structure.

CROSS HYBRIDIZATION OF DBH AND PNMT mRNA

Identification of DBH and PNMT clones was carried out by hybrid selection procedures detailed elsewhere[4]. While carrying out hybrid selection analysis with suspected DBH cDNA clones, the PNMT cDNA clone 22 was used as an internal control. Unexpectedly, this PNMT cDNA clone 22 appeared to hybrid select a messenger RNA which coded for a 72 kilodalton protein. Following immunoprecipitation with DBH antibody, it became apparent that PNMT clone 22 hybrid selected DBH mRNA in addition to PNMT mRNA[14]. Furthermore, DBH clones significantly cross hybrid select PNMT mRNA[14].

NORTHERN BLOT HYBRIDIZATION

^{32}P-labeled PNMT cDNA probe 22 and DBH probe 71 were used for hybridization with poly(A)mRNA bound to nitrocellulose. PNMT probe 22 hybridized to a mRNA from bovine adrenal medulla of approximately 1,100 nucleotides (nt) in length, using hybridization conditions described in the methods. Under slightly less stringent hybridization conditions, the PNMT probe hybridized weakly to a mRNA of unknown specificity having a length of 5,500 nt[14]. When DBH probe 103 was used as a probe in Northern blot hypridization, a mRNA of approximately 5,500 nt was hybridized, which was identical in size to the cross hybridizing mRNA species found when PNMT cDNA probe 22 was used under less stringent conditions.

SOUTHERN BLOT HYBRIDIZATION

High molecular weight DNA (greater than 30kb), isolated from bovine adrenal medulla was digested with various restriction enzymes and the digests dispersed on agarose gels and transferred to nitrocellulose. The filters were hybridized with DBH cDNA probe 71. It can be seen that the probe hybridized to 1-2 bands per lane. Comparison of DBH and PNMT Southern blots were shown in the same figure. PNMT cDNA probe 22 hybridizes to a 4.2 kb molecular weight band found with the DBH cDNA probe. In addition, PNMT 22 hybridized to a 14 kb Eco RI DNA band which is identical in MW to the top Eco RI band which hybridized to the DBH 71 probe.

POSSIBLE EVOLUTION OF CA SYNTHESIZING ENZYME

A large number of eukaryotic genes have recently been cloned[19], and studies of these cloned genes have shown the importance of duplication events in the evolution of individual genes and gene families[20] Homologies existing among these CA enzyme gene coding regions may be the result of duplication of a common ancestral gene, followed by divergent evolution with conservation of sequences essential for catalytic activity, substrate and cofactor binding, etc., to produce the unique enzyme functionalities.

The ancestral gene for the CA synthesizing enzyme is probably well conserved. This is based on the following evidence. Our antibodies directed against bovine adrenomedullary TH, purified by digestion of adrenomedullary protein with trypsin[5], have been used by many scientists around the world. These antibodies not only inhibit enzyme activity, but also localize TH in CA containing neurons and cells. The trypsin digested form of TH (tTH), having a 35 kilodalton molecular weight instead of the 60 kilodalton of the native form of TH, may possess not only the antigenic properties of TH, but also its enzymatically active site. It has been established that antibodies to tTH cross-react with the enzyme in many different animal species, including human, mouse, tortoise, and fish. The results suggest considerable conservation of amino acid sequence in the enzyme through evolution. Since portions of the TH molecule appear to have been conserved through evolution, duplication of a segment of this gene or an ancestral precursor could have resulted in the creation of similar gene coding sequences for the CA synthesizing enzymes. This probably indicates the presence of a gene family (a set of genes descended by duplication and variation from some ancestral gene) for the CA enzymes. The members of this gene family probably have related or identical functions, although they may be expressed at different times or in different cell types.

SPECIFIC IMMUNOLYSIS OF NERVE TERMINALS USING ANTISERA AGAINST NEUROTRANSMITTER SYNTHESIZING ENZYMES

We have recently observed (Docherty et al., submitted for publication) that incubation of striatal synaptosomes with antibodies to ChAT and complement apparently cause the antibody-dose dependent selective lysis of a cholinergic subpopulation of synaptosomes by blocking specific uptake of choline, and by releasing lactate dehydrogenase (LDH) and ChAT into the incubation medium. Antibodies to ChAT had no effect on the other subpopulations of the synaptosomes, such as GABA-ergic and dopaminergic synaptosomes. Similar experiments with antibodies to GAD caused selective lysis of the GABA-ergic subpopulation of the synaptosomes by releasing LDH and GAD into the medium, and blocking GABA uptake. Anti-TH also caused lysis of a dopaminergic subpopulation, but had no effect on the other cholinergic or GABA-ergic subpopulations.

The results indicate that antibodies to a neurotransmitter synthesizing enzyme selectively recognize a certain membrane protein or proteins of a specific subpopulation of synaptosomes which contain the neurotransmitter synthesizing enzyme, and suggest that neurotransmitter biosynthetic enzyme and a certain protein of the nerve endings share similar protein domain(s). This further suggests that the enzyme and the protein of the nerve endings also share common gene coding sequence(s).

The important implication of the above results is that the coordinate expression of the neurotransmitter enzyme and the nerve ending protein, may define the phenotype of the neuron. For DA neurons, coordinate expression of genes for TH and certain proteins of its nerve endings which are recognized by TH antibodies, is essential to determine the dopaminergic phenotype. The nerve ending protein which is recognized by TH-antibodies may be a partial or full size TH molecule, and this protein does not exist in other neuronal membranes, such as cholinergic and GABA-ergic membranes. Cholinergic phenotype may be decided by genes for ChAT, and ChAT-like protein in the nerve ending membranes. Similarly, coordinated expression of GAD and GABA-ergic neuronal membrane proteins may dictate the GABA-ergic phenotype.

SUMMARY

We found that the catecholamine biosynthetic enzymes, tyrosine hydroxylase, dopamine B-hydroxylase and phenylethanolamine N-methyltransferase share similar protein domains in their primary structures, and therefore are coded for by a single gene or a family of genes. In a recent report, we also demonstrated that antisera directed against tyrosine hydroxylase, choline acetyltransferase and glutamate decarboxylase cause specific complement-mediated lysis of dopaminergic, cholinergic and GABA-ergic subpopulation of synaptosomes, respectively. This implies that the neurotransmitter biosynthetic enzyme and the specific nerve ending protein(s) also share similar protein domain(s). Therefore, we postulate that the specific neurotransmitter biosynthetic enzyme and a certain membrane protein of the nerve endings probably share similar gene coding sequences and that coordinate expression of these proteins may determine the phenotype of the neuron.

REFERENCES

1. T.R. Elliot, The action of adrenaline, _J. Physiol._ (London) 32:401 (1905).

2. W.W. Douglas, Stimulus-secretion coupling: the concept and clues from chromaffin and other cells, _Brit. J. Pharmacol. Chemother._ 34:451 (1968).

3. R.M. Weinshelboum, N.B. Thoa, D.G. Johnson, I.J. Kopin, and J. Axelrod, Proportional release of norepinephrine and dopamine B-hydroxylase from sympathetic neurons, _Science_ 174:1349 (1971).

4. T.H. Joh and M. Goldstein, Isolation and characterization of multiple forms of phenylethanolamine N-methyltransferase, _Mol. Pharmacol._ 9:117 (1973).

5. T.H. Joh and M.E. Ross, Preparation of catecholamine-synthesizing enzymes as immunogens for immunohistochemistry, in: "Immunohistochemistry," vol. 3, A.C. Cuello, ed., John Wiley & Sons, Ltd., Chichester (1983).

6. T.H. Joh, D.H. Park, and D.J. Reis, Direct phosphorylation of brain tyrosine hydroxylase by cyclic AMP-dependent protein kinase: a mechanism of enzyme activation, _Proc. Natl. Acad. Sci. USA_ 75: 4744 (1978).

7. M.E. Ross, D.H. Park, G. Teitelman, V.M. Pickel, D.J. Reis, and

T.H. Joh, Immunohistochemical localization of choline acetyltransferase using a monoclonal antibody: a radioautographic method, Neuroscience 10(3):907 (1983).

8. E.E. Baetge, B.B. Kaplan, D.J. Reis, and T.H. Joh, Translation of tyrosine hydroxylase from Poly(A)mRNA in pheochromocytoma cells (PC12) is enhanced by dexamethasone, Proc. Natl. Acad. Sci. USA 78:1269 (1981).

9. T.H. Joh, C. Geghman, and D.J. Reis, Immunochemical demonstration in increased accumulation of tyrosine hydroxylase protein in sympathetic ganglia and adrenal medulla elicited by reserpine, Proc. Natl. Acad. Sci. USA 70:2767 (1973).

10. T. Lewander, T.H. Joh, and D.J. Reis, Tyrosine hydroxylase: delayed activation in central noradrenergic neurons and induction in adrenal medulla elicited by stimulation of central cholinergic receptors, J. Pharmacol. Exp. Ther. 200:523 (1977).

11. D.J. Reis, T.H. Joh, and R.A. Ross, Effects of reserpine on activities and amounts of tyrosine hydroxylase and dopamine-B-hydroxylase in catecholaminergic neuronal systems in rat brain, J. Pharmacol. Exp. Ther. 193:775 (1975).

12. R.A. Ross, T.H. Joh, and D.J. Reis, Increase in the relative rate of synthesis of dopamine-B-hydroxylase in the nucleus locus coeruleus elicited by reserpine, J. Neurochem. 31:1491 (1978).

13. R.A. Ross, T.H. Joh, and D.J. Reis, Reduced rate of biosynthesis of dopamine-B-hydroxylase in the nucleus locus coeruleus during the retrograde reaction, Brain Res. 160:174 (1979).

14. T.H. Joh, E.E. Baetge, M.E. Ross, and D.J. Reis, Evidence for the existence of homologous gene coding regions for the catecholamine biosynthetic enzymes, in: "Cold Spring Harbor Symposia on Quantitative Biology," vol. 48, J.D. Watson, ed., Cold Spring Harbor Laboratory, Cold Spring Harbor (1983).

15. B. Petrack, F. Sheppy, and V. Fetzer, Studies on tyrosine hydroxylase from bovine adrenal medulla, J. Biol. Chem. 243:743 (1968).

16. D.H. Park, E.E. Baetge, B.B. Kaplan, V.R. Albert, D.J. Reis, and T.H. Joh, Different forms of adrenal phanylethanolamine N-methyl-transferase: species-specific posttranslational modification, J. Neurochem. 38:410 (1982).

17. P.H. Seeburg, H. Shine, J.A. Martial, A. Urlich, J.D. Baxter, and H.M. Goodman, Nucleotide sequence of part of the gene for human chorionic somatomatropin: purification of DNA complementary to predominant mRNA species, Cell 12:157 (1977).

18. E.E. Baetge, H.M. Moon, B.B. Kaplan, D.H. Park, D.J. Reis, and T.H. Joh, Identification of clones containing DNA complementary to phenylethanolamine N-methyltransferase mRNA, Neurochem. Intl. 5(5):611 (1983).

19. J. Abelson, RNA processing and the intervening sequence problem, Ann. Rev. Biochem. 48:1035 (1979).

20. R. Breathnach and P. Chambon, Organization and expression of Eucaryotic split genes coding for proteins, Ann. Rev. Biochem. 50:349 (1981).

MOLECULAR STUDIES OF PITUITARY GLAND DIFFERENTIATION

John E. Pintar

Department of Anatomy and Cell Biology
Columbia University, College of Physicians and Surgeons
630 West 168th Street, New York, NY   10032

INTRODUCTION

All hormone-synthesizing cells in the pituitary gland appear
to differentiate from an initially homogeneous epithelium called
Rathke's pouch, which forms as an invagination of the oral ectoderm (Levy et al., 1980). Both the anterior and intermediate
lobes of the pituitary are derived from this structure, which
separates from the oral ectoderm during early embryogenesis. The
separation occurs after Rathke's pouch has contacted a ventral
outpocketing of the neural tube, the infundibulum, which later
becomes the posterior pituitary. The anterior wall of Rathke's
pouch then proliferates rapidly and becomes the anterior pituitary, which contains at least six different cell types synthesizing different peptide hormones; cells producing proopiomelanocortin (POMC) account for about 5% of the cells in the adult
anterior lobe. The posterior wall of Rathke's pouch differentiates
into the intermediate lobe which contains only cells synthesizing
POMC; however, these POMC-containing cells differ from anterior
lobe POMC cells in that they process this prohormone to different
end-product hormones (see below).

A variety of indirect evidence has suggested that differentiation of the intermediate lobe is dependent on close contact
between the posterior wall of Rathke's pouch and the infundibulum.
This evidence has resulted from observations on cocultures of
anterior and posterior lobes, from _in vivo_ studies in which the
infundibulum is lacking or removed, and from observations in
species lacking an intermediate lobe (e.g. avians and cetaceans),
where interaction between Rathke's pouch and the infundibulum is

interrupted by intervening mesenchymal tissue (Gaillard, 1937; Hanaoka, 1967; Wingstand, 1951, 1966). Thus an important research objective is to understand at a biochemical level the influence of the posterior lobe or its precursor tissue, the infundibulum, on Rathke's pouch differentiation.

In this chapter we will summarize our recent studies of POMC gene expression and processing and monoamine oxidase activity during pituitary development and discuss whether these may be useful as markers for intermediate lobe differentiation. In addition, we will present evidence that the extracellular matrix surrounding different regions of the fetal pituitary differs even at early stages of development and thus may contribute to lobe-specifc differentiation.

CHARACTERIZATION OF POMC-RELATED PEPTIDES IN THE PRENATAL RAT PITUITARY GLAND

A number of pituitary peptide hormones including ACTH, B-endorphin (B-EP), B-lipotropin (B-LPH), a-MSH and β-MSH are derived from a common precursor (POMC; see Eipper and Mains, 1980 for review). Different processing endpoints of a similar and probably identical POMC molecule in anterior and intermediate lobe POMC-producing cells give rise to different hormone endproducts that have quite diverse physiological effects. The sequences of processing events and the description of POMC molecular inter-mediates have recently been described in detail for adult rat and mouse POMC cells in both lobes. In intermediate lobe cells, POMC-derived peptides that are processing endpoints in anterior lobe cells (B-LPH and ACTH) are proteolytically cleaved to smaller molecular weight peptides. In addition, some of these peptides (ACTH 1-13) and B-endorphin (1-31)) are further modified by acety-lation, amidation, and further proteolytic cleavages to produce a-MSH and modified forms of endorphin. These modifications appear to be confined to POMC producing cells in the intermediate lobe of the adult rat (Liotta et al., 1980; see Zakarian and Smyth, 1982, for review).

Since all recent evidence indicates that only one POMC gene is present in the rat (Eberwine and Roberts, 1983) it seems likely that lobe-specific processing differences are regulated by the presence of the processing enzyme themselves. While there has been recent progress in the isolation of secretory granules where these processing conversions occur and in characterization of processing enzymes (Gumbiner et al., 1981; Loh et al., 1982; Glembotski, 1982a, 1982b; Powers and Nasjilletti, 1982), it is not known whether there are developmental changes in the time of appearance of the multiple processing enzymes required to produce all peptide forms known to occur in adults.

Numerous immunocytochemical demonstrations of POMC-related peptides in fetal rat pituitary have been reported (Begeot et al., 1977; Chatelain et al., 1976, 1979; Catelain and Dupuoy, 1981; Coffigny et al., 1978; Watanabe and Daikoku, 1979; Osumora and Nakane, 1982; Schwartzenberg and Nakane, 1982, Khachaturian et al., 1983). In most cases, these studies cannot reveal the extent of POMC processing in the cells containing immunoreactive material because the immunocytochemical cross reactivity of antibodies made to POMC end-product hormones,to the POMC precursor and processing intermediates cannot be determined. This uncertainty may account for the range in the reported initial appearance of the POMC-related peptides during fetal development (Dupuoy, 1980).

For these reasons, recent studies have begun to examine the molecular forms of POMC-related peptides present in the fetal pituitary glands of a number of species including sheep and human (Loh et al., 1980; Silman et al., 1976, 1978; Brubaker et al., 1982). The results of these studies have not been consistent, however, and most have used whole fetal pituitaries. Thus it has been impossible to determine whether cells in specific lobes are initiating POMC synthesis and lobe-specific processing prior to cells in the other lobe and whether processing patterns in specific lobes change during embryogenesis.

Thus we have begun a detailed analysis of the molecular forms of POMC-derived peptides present prenatally in the rat embryo in order to provide information on numerous related problems. These analyses should be able to demonstrate when lobe-specific processing patterns are established during embryogenesis, when particular processing enzyme activities become functional, and should clarify previous immunocytochemical studies.

In initial experiments, we sought to determine whether the major adult lobe-specific processing patterns for POMC are established prenatally in the rat (Allen et al., 1984). We utilized the fact that the high molecular weight POMC precursor, its major processing intermediates, and many peptide end-products of both lobes can be separated by SDS-polyacrylamide gel electrophoresis under denaturing conditions; adult anterior lobe extracts are characterized by the predominant presence of immunoreactive B-LPH and ACTH (1-39), while adult intermediate lobe cell extracts are characterized by the presence of B-endorphin, CLIP, a-MSH, and an absence of B-LPH and ACTH. It should be noted that B-EP-related peptides B-EP (1-31), B-EP (1-27), B-EP (1-26) and the N-acety-lated forms of these peptides) cannot be separated by this gel system and must be resolved by ion-exchange chromatography and HPLC (see below).

We initially characterized the distribution of molecular weight forms of ACTH and B-EP-related peptides in separated anterior and neurointermediate lobes of fetal rat pituitaries from late gestational ages (embryonic days 18 and 20, e18, e20; gestation period 22 days). Fetal pituitary extracts were prepared and separated via SDS-electrophoresis; the slices were eluted and assayed for ACTH and B-EP radioimmunoreactivity using both an ACTH antiserum that recognizes the 11-20 region of ACTH and all POMC molecular intermediates containing this region and a B-EP antiserum that recognizes B-EP, B-LPH, and POMC.

Isolated anterior lobe extracts from both e18 and e20 fetuses showed a distribution of POMC-related peptides similar to that seen in the anterior lobe of adults; most ACTH immunoreactive material in the anterior lobe comigrated with ACTH (1-39), while approximately equal amounts of B-EP and B-LPH were present. In the isolated neurointermediate lobes, on the other hand, the major POMC-related peptide present was B-EP, with very little 4.5 K ACTH and no B-LPH; again, these patterns are similar to adult intermediate lobes. Taken together, these patterns strongly suggest that major aspects of lobe-specific POMC proteolytic processing are established before birth in the rat.

Extracts of whole fetal pituitaries from earlier ages (e14 and e16) were then analyzed for the presence of POMC-related peptides using the above procedures. At these ages the predominant peptide found was B-EP with relatively small amounts of B-LPH and ACTH present (Allen et al., 1984; see Figure 1a for day e17 profile). Thus fetal pituitaries from these ages showed POMC peptide profiles similar to those observed in the adult intermediate lobe. There are two possible explanations for these results. First, it is possible that POMC synthesis first occurs in the intermediate lobe. Although this possibility is not supported by most immunocytochemical studies, it may be that the amounts of POMC-related peptides present at these ages are below the levels required for light immunocytochemical demonstration. Other studies have shown that B-lutenizing hormone radioimmuneactivity is present earlier than light immunocytochemical demonstrations of this peptide (Nemeskeri et al., 1983). Alternatively, it is possible that anterior lobe POMC cells contain these peptides and alter the extent of proteolytic processing at a subsequent stage of development. In situ hybridization studies that can identify specific cells containing POMC mRNA (see below) should be able to resolve this question.

Assays for a-MSH using antisera that recognize the acetylated end of this molecule did not detect a-MSH until day 19 of development (Allen et al., 1984). This raised the question of whether ACTH (1-13) was present but non-acetylated at earlier ages and whether B-EP immunoreactive material present at early fetal

ages included modified B-EP peptides in addition to B-EP (1-31).
In initial studies addressing these questions, we have analyzed
extracts of e15-postnatal day 6 (p6) rat pituitary glands for
acetylated, non-acetylated, and trimmed forms of B-endorphin
(B-EP) and melanocyte stimulating hormone (MSH) using specific
antisera to assay fractions isolated by successive use of gel
filtration, ion exchange chromatography, and HPLC (Pintar et al.,
1984).

The predominant, and possibly only, form of MSH present in
day e15-e17 extracts is desacetyl-a-MSH, and the only form of B-EP
detected is B-EP(1-31) (see Figure 1). Acetylated B-EP(1-31)
and acetylated and non-acetylated B-EP(1-26) and B-EP(1-27) are

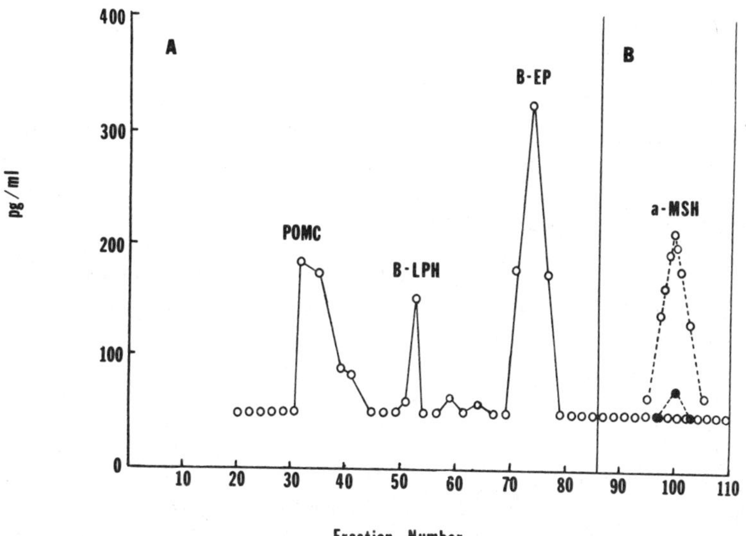

Figure 1.   Sephadex G-50 analysis of endorphin and MSH immuno-
             reactivity in e17 rat pituitary. Embryonic pituitary
             extracts were eluted in acetic acid and fractions
             assayed with both an endorphin antiserum recognizing
             POMC, B-LPH, and B-EP (Figure 1a) and an antiserum
             recognizing only acetylated forms of B-EP; no fractions
             assayed with the acetylation-requiring antiserum gave
             detectable radioimmune activity at this age. Later-
             eluting fractions were assayed with antisera to a-MSH
             (Figure 1b) that recognize des-acetyl-a-MSH.((ACTH
             1-13);open circles) or exhibit limited cross-reactivity
             (0.2%) with this peptide but do recognize acetylated
             ACTH (1-13) (closed circles). The results show that
             essentially all MSH-sized immunoreactive material is
             present as des-acetyl MSH (ACTH 1-13).

not detected prior to day e18. Since acetylated and trimmed B-EP
forms are present only in adult intermediate lobe, e21 and p6
anterior and neurointermediate lobes were separated before extraction and analyzed separately by ion-exchange chromatography and
HPLC to determine when and in which lobes these peptides first
appeared. At e21, acetylated and shortened B-EP forms are present
only in the neurointermediate lobe. Acetylated-B-EP(1-31) is
the predominant form at this age; of the trimmed B-EP forms
present, >90% of both B-EP(1-26) and B-EP(1-27) are acetylated;
this suggests that acetylation of these peptides occurs before
trimming. However, even at p6, the relative abundance of acetylated B-EP(1-26) and acetylated-B-EP(1-27) is still lower than
that found in adult intermediate lobe. The forms of MSH present
at these later stages were also analyzed. At e21, >90% a-MSH in
extracts of the neurointermediate lobe is acetylated, while only
25% is di-acetylated; interestingly, desacetyl-a MSH is still
present in the anterior lobe both at this stage and at p6.

These results demonstrate a temporal dissociation during
development between production, acetylation, and trimming of B-EP
and production and acetylation of a-MSH. Ontogeny of these modifications parallels the biosynthetic pathway established in adult
intermediate lobe. Since acetylated and non-acetylated peptides
have different biological activities, these observations should
provide a basis for assessing the role of these peptides in
development.

POMC GENE EXPRESSION DURING DEVELOPMENT

The initial time that immunological demonstration of specific
gene products occurs may not reflect the times that genes are
first activated during development; mRNA production may precede
significant mRNA translation or peptide storage. For example, the
casein gene family is transcribed throughout pregnancy, but is not
translated in abundance until near birth (Rosen et al., 1975). Of
direct relevance to our studies is the observation that cultured
lines of cells derived from Rathke's pouch, although lacking
secretory granules and unable to store peptides, are able to
produce and release to the medium a variety of pituitary hormones
in culture (Ishikawa et al., 1977; Shiino et al., 1977). If such
were the case in vivo, then these cells might be actively synthesizing hormones at early stages of development, but not storing
sufficient quantities to ensure immunologic demonstration. Thus
it is important to have an independent measure of gene activity.
Since cDNA probes for POMC have been isolated, we have begun using
these probes to determine whether the first detections of POMC
gene expression and peptide synthesis coincide and whether there
are significant changes in POMC mRNA levels that can be correlated
with specific developmental events.

We have initiated studies using POMC cDNA probes to determine
the earliest stage at which POMC mRNA synthesis is detectable
during development (Pintar et al., 1982). We have analyzed
nucleic acid extracts of different embryonic tissues at various
stages of development with labeled cDNA probes to POMC mRNA. We
have detected POMC mRNA by a modified "dot-blot" procedure
in which isolated nucleic acid is spotted and baked onto nitrocellulose and then probed with radiolabeled POMC cDNA under
stringent conditions; bound cDNA is then revealed by autoradiography of X-ray film. We have thus far been able to detect and
quantitate POMC mRNA in embryonic pituitaries at stages when POMC
peptides are known to be present. In initial experiments, aliquots of isolated nucleic acid from various el5 rat embryonic
tissues were spotted onto nitrocellulose and hybridized with
^{32}P-POMC cDNA. We have reproducibly detected specific POMC cDNA
binding when 1/5 total nucleic acid from one el5 pituitary was
analyzed; this results strongly suggests that sufficient POMC mRNA
is present at this stage to allow us to locate POMC mRNA-producing
cells by <u>in situ</u> hybridization, and clarify the site of synthesis
of "intermediate-like" POMC synthesis at this stage (see above).

In addition, we have begun to compare the relative amount of
POMC mRNA present at different ages (Pintar et al., 1982). The
amounts of radiolabelled probe bound to different extracts on
nitrocellulose filters have been quantitated by scanning densito-
metry of autoradiograms and expressed as the ratio of density of
the radiolabeled spot to the amount of nucleic acid spotted. Both
the densities of the autoradiographic spots (revealing POMC probe
binding) and ethidium bromide-staining of sample aliquots spotted
onto agar plates (for nucleic acid determintion) were within the
linear ranges of standard curves. Our initial results showed that
the amount of POMC mRNA per microgram of nucleic acid does not
change between embryonic day 15 and embryonic day 19. The most
likely explantation for this result is that the proportion of POMC
cells and the amount of POMC mRNA/cell is remaining constant
during this time period; we expect that analysis of tissue sections by <u>in situ</u> hybridization will enable us to confirm this. We
have also determined that the ratio of intermediate to anterior
lobe POMC mRNA per microgram of nucleic acid in the neonate (2:1)
has begun to increase toward the adult ratios (10:1) from amounts
found in el9 extracts (1:1), suggesting that at or near birth
greater intermediate lobe POMC gene activity or message stabilization is occuring.

Quantitation of POMC mRNA production in developing pituitaries
does not provide information about the spatial distribution of
POMC-producing cells or whether the proportion of cells making
POMC mRNA changes during development. Use of radiolabeled cDNA as
histochemical reagent will allow us to analyze mRNA production at
the single-cell level in tissue sections by <u>in situ</u> hybridization

(Brachic et al., 1982; Gee et al., 1983). In initial studies we
have hybridized ^{32}P-labelled probes to tissue sections of e21
pituitaries. At this age, the intermediate lobe region of the
tissue section has bound more probe than the adjacent anterior
lobe, which is consistent with the above observation that the
amount of POMC mRNA in the intermediate lobe is beginning to
increase at this time.

DISTRIBUTION OF MONOAMINE OXIDASE B IMMUNOREACTIVITY IN ADULT AND EMBRYONIC RAT PITUITARY

In addition to pituitary peptide hormones, other lobe-specific
gene products may be expressed during development of the pituitary
gland; one such possibility that may have important physiological
consequences is the enzyme monoamine oxidase (MAO), which is the
primary enzyme involved in degradation of biogenic amines such as
norepinephrine, serotonin, and dopamine (see Pintar and Breakefield, 1982 for review); these amines directly influence the
synthesis and release of a number of pituitary peptide hormones.

We and others have recently established that two pharmacolog-
ically defined MAO activities (MAO-A and MAO-B) are mediated by
different proteins. We have made antibodies to the MAO-B protein
which do not cross-react with MAO-A (Pintar et al., 1983) and
which have been used to localize MAO-B immunoreactivity in the rat
central nervous system (Levitt et al., 1982).

We initially used MAO-B specific antisera to localize MAO-B
immunoreactivity in the adult pituitary and have made three main
observations (Figure 2; Cooper et al., 1983). First MAO-B immunoreactivity is found in the major cell class of the posterior lobe,
the pituicytes; this localization is consistent with other recent
observations that these cells express other astrocyte-like properties (Suess and Pliska , 1981; Salm et al., 1982). Since astrocytes have the ability to internalize exogenous amines (Tardy et
al., 1982), this localization of MAO activity indicates that these
cells may inactivate transmitters that are released from hypothalamic nerve endings in the posterior lobe. Second, MAO-B
immunoreactivity is found in a "boundary" layer of cells between
the anterior and intermediate lobes. Although these cells had
previously been morphologically described, no discrete biochemical
characteristic has previously been ascribed to them. This localization indicates that these cells may insulate anterior and
intermediate lobes from physiological effectors released in the
other lobe. Third, selected scattered cells in the anterior lobe
(about 10% of the total population) expressed MAO-B immunoreactivity; no staining was observed in the intermediate lobe.

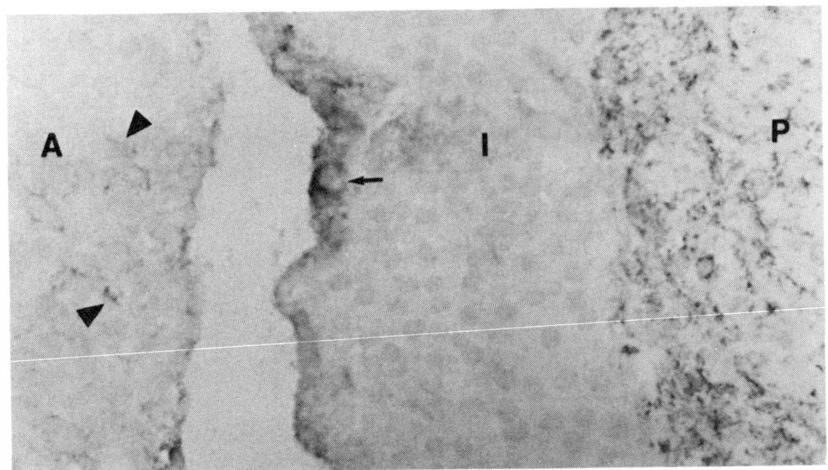

Figure 2. MAO-B immunohistochemcial staining of adult rat pituitary. The pituicytes in the posterior lobe (P) are
stained and the staining of the marginal layer of cells
bordering the pituitary cleft is apparent (arrow). In
addition, a small percentage of cells in the anterior
lobe (A) are also stained.(arrowheads).

Our recent studies indicate that these three cell populations
accumulate detectable MAO-B immunoreactivity at different times
during development (Pintar and Levitt, in preparation). MAO-B
immunoreactiivty is first seen in pituicytes at e16, the earliest
stage examined thus far; by birth most pituiticytes were MAO-B
positive. On the other hand, MAO-B immunoreactivity was not
detected in marginal cells until e21; by the end of the first
postnatal week, most marginal cells are MAO-B positive. Finally
MAO-B immunoreactivity remains undetectable in anterior lobe cells
even at the end of the second post-natal week. These differences
in the time of appearance of MAO-B positive populations suggests
that different regulatory functions for this enzyme may be activated at different times during development.

CHARACTERIZATION OF THE EXTRACELLULAR MATRIX DURING PITUITARY
DEVELOPMENT

As mentioned above, the interaction between Rathke's pouch
and the infundibulum is established during early development and
is maintained in species that have an intermediate lobe; experimental interference with this interaction leads to impaired
intermediate differentiation.

59

There is suggestive evidence that the extracellular matrix (ECM) may have a direct role in pituitary gland differentiation (Fremont and Ferrand, 1979; Watanabe 1982; Beugot et al., 1982). Under normal conditions in vivo, the first cells of Rathke's pouch to differentiate are those near the basement membrane lining the basal surface of the pouch. In vitro, however, the first differentiating cells are near the cavity of Rathke's pouch (Fremont and Ferrand, 1979). In these experiments, the basement membrane had been enzymatically removed and the highest concentration of matrix in vitro was within the cavity, exactly reversed from the in vivo situation. The nature of these cytologically differentiated cells was not investigated, nor was the nature of the matrix elucidated.

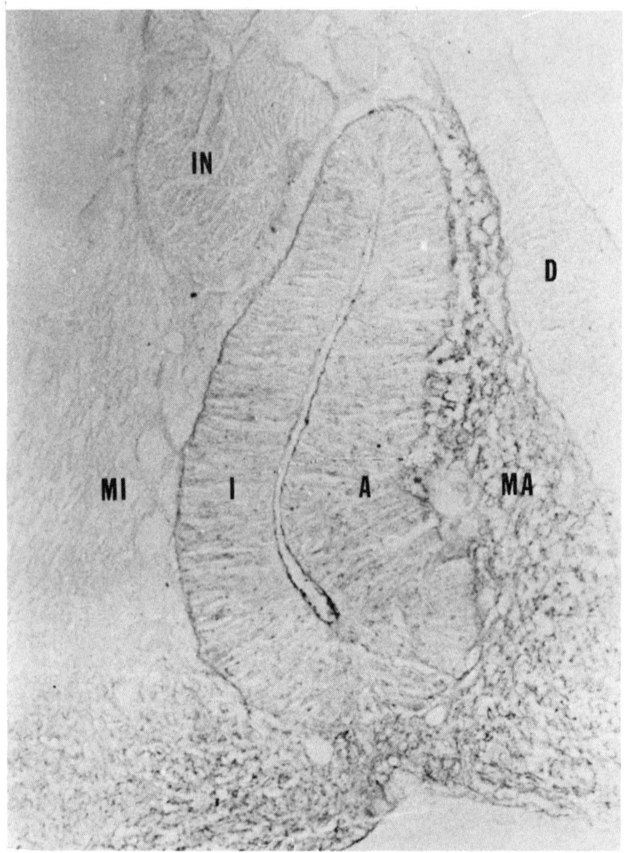

Figure 3.   Sagittal section of newly formed e14 rat pituitary primordium. The tissue was fixed with CPC-formaldehyde and stained with Alcian blue under conditions that stain only extracellular glycosaminoglycans (GAG; Derby and Pintar, 1978) Mesenchymal cells (MA) adjacent to the presumptive anterior lobe (A) are surrounded by a GAG-rich environment. In contrast, the
(continued)

Figure 3 (continued)
> environment near the presumptive intermediate lobe
> (I) (which includes the infundibulum (IN), associated
> mesenchyme (MI), and extracellular matrix) is GAG-poor.
> The floor of the diencephalon is denoted by D.

We have begun to examine extracellular matrix materials
present in the region of the developing rat pituitary _in vivo_ and
have initally focused on extracellular glycosaminoglycans (GAG) in
this region. Embryos (e14) were fixed in fixative containing
cetylpyridinium chloride to maximally preserve water soluble GAG
(Derby and Pintar, 1978) and stained for GAG with alcian blue.
Alcian blue staining is present in the basal lamina of the
newly formed pituitary (Figure 3); in addition, significantly more
staining is associated with the mesenchyme adjacent to the presumptive anterior lobe than with the mesenchyme and infindibulum
associated with the intermediate lobe. This initial result
suggests that a further examination of other matrix material
present in this region is warranted and supports the idea that
differences in matrix may be important in pituitary differentiation.

The author wishes to thank G. Andrew Stover for photographic
assistance. This work was supported in part by ND-18592.

REFERENCES

Allen, R.G., Pintar, J.E., Stack, J., and Kendall, J. (1984).
Biosynthesis and processing of pro-opiomelanocortin-derived
peptides during fetal pituitary development. _Dev.Biol_. 102:
43-50.

Begeot, M., Dubois, M.P. and Dubois, P.M. (1982). Comparative
study _in vivo_ and _in vitro_ of the differentiation of immunoreactive corticotropic cells in fetal rat anterior pituitary.
Neuroendocrinology 35:255-264.

Begeot, M., Dubois, M.P., and Dubois, P.M. (1977). Immunological
localization of alpha-endorphins and beta-lipotropin in
corticotropic cells of normal and anencephalic fetal pituitaries. _Cell Tiss. Res._ 193:413-422.

Brubaker, P.L., Baird, A.C., Bennett, H.P.J., Browne, C.A. and
Solomon, S. (1982). Corticotropic peptides in the human fetal
pituitary. _Endocrinology_ 111:1150-1159.

Chatelain, A., Dubois, M.P. and Dupuoy, J.P. (1976). Hypothalamus
and cytodifferentiation of fetal pituitary-gland study, in
vivo. _Cell Tiss. Res._ 169:335-344.

Chatelain, A., Dupouy, J.P. and Dubois, M.P. (1979). Ontogenesis
of cells producing polypeptide hormones ACTH, MSH, LPH, GH,
prolactin) in the fetal hypophysis of the rat: influence of
the hypothalamus. _Cell Tiss. Res._ 196:409-427.

Chatelain, A. and Dupouy, J.P. (1981) Adrenocorticotrophic hormone in the anterior and neurointermediate lobes of the fetal rat pituitary gland. J. Endocrinol. (Lond.) 89:181-186.

Coffigny, H., and Dupouy, H.P. (1978). Fetal adrenals of rat-correlations between growth, cytology, and hormone activity, with and without ACTH deficiency. Gen. Com. Endocrinol. 34:321-322.

Cooper, V. , Pintar, J.E., and Levitt, P. (1983). Localization of monoamine oxidase B immunoreactivity in the neonate and adult rat pituitary gland. Neuroscience Abst. 9:704.

Derby, M.A. and Pintar, J.E. (1978). The histochemical specificity of Streptomyces hyaluronidase and chondroitinase ABC. Histochem J. 10:529-547.

Dupouy, J.P. (1980) Differentiation of MSH-containing, ACTH-containing, endorphin-containing and LPH-containing cells in the hypophysis during embryonic and fetal development. Int. Rev. Cytol. 68:197-249.

Eberwine, J.H. and Roberts, J.L. (1983). Analysis of pro-opiomelanocortin gene structure and function. DNA 2:1-8.

Eipper, B.A. and Mains, R.E. (1980). Structure and biosynthesis of pro-adrenocorticotropin/endorphin and related peptides. Endocrine Rev. 1:1-27.

Fremont, P.H. and Ferrand, R. (1979). In vitro studies on the self-differentiating capacities of the quail adenohypophysis epithelium. Anat. Embryol. 156:255-270.

Gaillard, P.J. (1937). An experimental contribution to the origin of the pars intermedia of the hypophysis. Acta Neurol. Morphol. 1:3-11.

Gee, C., Chen, C.-L., Roberts, J.L., Thompson, R. and Watson, S.J. (1983). Identification of proopiomelanocortin neurones in rat hypothalamus by in situ cDNA-mRNA hybridization. Nature 306:374-376.

Glembotski, C. (1982a). Characterization of the peptide acetyltransferase activity in bovine and rat intermediate pituitaries responsible for the acetylation of B-endorphin and a-melanotropin J. Biol. Chem. 257:10501-10509.

Glembotski, C. (1982b) Acetylation of a-Melanotropin and B-endorphin in the rat intermediate pituitary. J. Biol. Chem. 257: 10493-10500.

Gumbiner, B. and Kelley, R. (1981). Secretory granules at an anterior pituitary cell line, A & T-20 contain only mature forms of corticotropin and B-lipotropin. Proc. Natl. Acad. Sci. USA 78:318-322.

Haase, A.T., Stowring, L., Harris, J.D., Traynor, B., Ventura, P., Peluso, R. and Brahic, M. (1982). Visual DNA synthesis and the tempo of infection in vitro. Virology 119:399-410.

Hanaoka, Y. (1967). The effects of posterior hypothalectomy upon the growth and metamorphosis of the tadpole of Rana pipiens. Gen. Comp. Endo. 8:417-431.

Ishikawa, H., Shiino, M., and Rennels, E.G. (1977). Functional clones of pituitary cells derived from Rathke's pouch epithelium of fetal rats. Endocrinology 100:1227-1230.

Khachaturian, H., Alessi, N.E., Munfakh,N., and Watson, S.J. (1983). Ontogeny of opiod and related peptides in the rat CNS and pituitary. Life Sci. 33:61-64.

Liotta, A.S., Yamaguchi, H. and Krieger, D.T. (1981). Biosynthesis and release of B-endorphine-, N-acetyl B-endorphine, B-endorphin-(1-27)-, and N-acetyl B-endorphin-(1-27)-like peptides by rat pituitary neurointermediate lobe: B-endorphin is not further processed by anterior lobe. J. Neurosci. 1:585-595.

Levitt, P., Pintar, J.E. and Breakefield, X.O. (1982). Monoamine oxidase B is found in brain astrocytes and serotonergic neurons. Proc. Natl. Acad. Sci. USA, 79:6385-6389.

Levy, N.B., Andrew, A., Rawdon, B.B. and Kramer, B. (1980). Is there a ventral neural ridge in chick embryos? Implications for the origin of adenohypophyseal and other APUD cells. J. Embryol. exp. Morph. 57:71-78.

Loh, Y.P., Eskay, R.L. and Brownstein, M. (1980). MSH-like peptides in rat brain: identification and changes in level during development. Biochem. Biophys. Res. Comm. 94:916-923.

Loh, Y.P., Gritsch, H.M., and Chang, T.-L. (1982). Pro-opiomelanocortin processing in the pituitary: A model for neuropeptide biosynthesis. Peptides 3:397-404.

Nemeskeri, A., Halasz B., and Kurcz, M. (1983). Ontogenesis of the rat hypothalamo-adenohypophyseal system and the capacity of the fetal pituitary to differentiate into hormone-synthesizing and releasing cells. In The Anterior Pituitary Gland, ed. Bhatnager, Raven.

Osamura, R.K. and Nakane, P.K. (1982). Review: Functional Differentiation of cells in the anterior and intermediate pituitary glands-immunohistochemical studies. Acta Histochem. Cytochem. 15:294-309.

Pintar, J.E. and Breakefield, X.O. (1982). MAO activity as a determinant in human neurophysiology. Behav. Genet. 12:53-68.

Pintar, J.E., Roberts, J.L. and Gee, C. (1982). Expression of the propiomelanocortin gene in early stages of fetal rat pituitary development. Neurosci. Abst. 8:703.

Pintar, J.E., Levitt, P., Salach, J.I., Weyler, W., Rosenberg, M.B., and Breakefield, X.O. (1983). Specificity of antisera prepared against pure bovine MAO-B. Brain Res. 276:127-139.

Pintar, J.E., Kreiger, D.T., and Liotta, A.S. (1984). Ontogeny of endorphin and MSH-sized peptides during rat pituitary develoment. Int. Congress. Endo. 7, in press.

Powers, C.A. and Nasjilletti,A, (1982). A novel kinin-generating protease (Kininogenase) in the porcine anterior pituitary. J. Biol Chem. 257:5594-5600.

Rosen, J.N., Wou, S.L.C. and Comstock, J.P. (1975). Regulation of casein messenger RNA during development of the rat mammary gland. Biochem. 14:2895-2903.

Salm, A.K., Hatton, G.I. and Nilaver, G. (1982). Immunoreactive glial fibrillary acidic protein in pituicytes of the rat neurohypophysis. Brain Res. 236:471-476.

Shiino, M., Ishikawa, H. and Rennels, E.G. (1977). In vitro and in vivo studies on cytodifferentiation of pituitary clonal cells derived from the epithelium of Rathke's pouch. Cell Tiss. Res. 181:473-485.

Silman, R.E., Chard, T., Lowry, P.J., Smith, I. and Young, I.M. (1976). Human fetal pituitary peptides and parturition. Nature 260:716-718.

Silman, R.E., Holland, D., Chard, T., Lowry, P.J., Hope, J., Robinson, J.S. and Thornborn, G.D. (1978). The ACTH "family" tree of the monkey changes with development. Nature 276:526-527.

Suess, U. and Pliska, V. (1981). Identification of the pituicytes as astroglial cells by indirect immunofluorescence-staining for the glial fibrillary acidic protein. Brain Res. 221:27-33.

Schwartenberg, D.G., and Nakane, P.K. (1982). Ontogenesis of adrenocortico-tropin related peptide determinants in the hypothalamus and pituitary gland of the rat. Endocrinology 110:855-864.

Tardy, M., Costa, M.F.D., Fages, C., Bardakdjian, J. and Gonnard, P. (1982). Uptake and binding of serotonin by primary cultures of mouse astrocytes. Devel. Neurosci. 5:19-26.

Watanabe, Y.G. (1982). An organ culture study on the site of determination of ACTH and LH cells in the rat adenohypophysis. Cell Tiss. Res. 227:267-276.

Watanabe, Y.G. and Daikoku, S. (1979). An immunohistochemical study on the cytogenesis of adenohypophysial cells in fetal rat. Devel. Biol. 68:557-567.

Wingstrand, K.G. (1951). The structure and development of the avian pituitary. Lund: Gleerup.

Wingstrand, K.G. (1966). Comparative anatomy and evolution of the hypophysis. In The pituitary gland (ed. G.W. Harris and B.T. Donovan), vol. III, p. 1-27, London: Butterworths.

Zakarian, S. and Smyth, D.G. Distribution of B-endorphin-related peptides in rat pituitary and brain. Biochem J. 202:561-571.

NOVEL INSULIN-RELATED SEQUENCES IN FETAL BRAIN

Lydia Villa-Komaroff, Antonio Gonzalez,
Song Hou-Yan, Bruce Wentworth and Paul Dobner

Molecular Genetics and Microbiology
University of Massachusetts Medical Center
Worcester, MA 01605

INTRODUCTION

An examination of the literature leaves no doubt that there is
a moiety which cross-reacts with anti-insulin antibodies in the
central and peripheral nervous systems. However, there is
considerable doubt as to how much of this material there is, where
exactly it resides, where it originates and what role it plays. In
this chapter, we will review the evidence for the presence and
synthesis of insulin or insulin-like peptides in the nervous system
and we will describe how we are using recombinant DNA technology to
identify and characterize RNA sequences homologous, but not
identical, to sequences encoding insulin. (We will not deal with
immunoreactive insulin in invertebrates [for review, see Le Roith
et al].)

INSULIN IN THE CENTRAL AND PERIPHERAL NERVOUS SYSTEM

Because of insulin's role in glucose metabolism and since the
brain uses large amounts of glucose, it is not unreasonable to
suppose insulin might play a role in brain metabolism. Although it
is clear that fluctuations in peripheral insulin levels do not
affect glucose metabolism in the CNS in an acute fashion, the
effect of plasma concentrations of insulin on the levels of insulin
in the brain and on brain metabolism is disputed. If insulin is
injected into the jugular vein, there is no increase in glucose
uptake in rat brain, as measured by the uptake of tritiated
2-deoxyglucose, in the ventral medial, ventral lateral, or dorsal
hypothalamus or in an area from the cerebral cortex (Goodner et al,
1980). In man, infusion of both insulin and glucose appears to

increase the transport of glucose across the blood brain barrier,
but not the net uptake of glucose by the brain (Hertz et al,
1981). The intraperitoneal injection of a subconvulsive dose of
insulin changes monoamine metabolism in the brain (DeMontis et al,
1978; MacKenzie and Trulson, 1978; Berggren et al, 1983).

The levels of insulin in the brain and in the cerebral spinal
fluid (CSF) have been measured under conditions where plasma
insulin levels are elevated by infusion, injection or by
stimulating the endogenous production of insulin (Woods and Porte,
1977; Ratzmann and Hampel, 1980; Yallow and Eng, 1983; Margolis and
Altszuler, 1967) and under conditions where insulin levels are
decreased by streptozotocin treatment which destroys the pancreatic
B cells, (Havrankova et al, 1979, Oomura and Kitu, 1981). There is
general agreement that levels of insulin in the CNS and CSF do not
respond acutely to serum levels of insulin. However, CSF levels of
insulin do appear to be related to serum levels, although
equilibrium is reached slowly and does not fluctuate as much as
plasma levels do (Woods and Porte, 1977). Some investigators
(Yallow and Eng, 1983) report that the level of insulin in the
brain is a reflection of plasma levels and others report that brain
levels of insulin are independent of plasma levels (Havrankova et
al, 1979: Oomura and Kita, 1981). One report postulates that the
permeability of the blood-brain barrier is affected in diabetic
humans (Lorenzi, et al, 1980).

The brain could obtain insulin from the circulation and still
be protected from changes in the serum level if insulin entered the
CNS in regions of the brain where there is no blood brain barrier
and was internalized and sequestered. This does happen to some
extent; labelled insulin, injected into the circulation, binds to
nerve terminals in regions of the hypothalamus where there is no
blood brain barrier, especially the external median eminence and
the arcuate nucleus (van Houten and Posner, 1981; van Houten, et
al, 1980). Whether this insulin is further internalized is not
clear since the brains were examined five minutes after injection
of the labeled insulin. It would be very useful to follow the fate
of this insulin after longer times.

In addition to the insulin-binding sites identified directly by
injecting ^{125}I-insulin and using autoradiography (van Houten et
al, 1979; van Houten and Posner, 1981), insulin receptors can be
detected in all parts of the CNS (Havrankova and Roth, 1978; Pacold
and Blackard, 1979). Insulin receptors can also be found on blood
vessels throughout the CNS (van Houten and Posner, 1979). This
raises the possibility that all of the insulin and insulin
receptors detected in the CNS are associated with the endothelial
layer of cells in the microcapillaries which constitute the
blood-brain barrier. Goodner and Berrie (1977) provided the first
evidence that insulin does not cross the blood-brain barrier. In

vitro studies indicate that isolated microvessels contain insulin
receptors (Frank and Partridge, 1983). The possibility remains
that insulin could affect brain metabolism by interacting with the
receptors in the barrier. A rough estimate of how much CNS insulin
could be accounted for by binding at the capillary endothelium of
the blood-brain barrier falls about midway in the range of values
reported (Frank and Partridge, 1983).

The finding that insulin has a marked and measurable effect
when injected directly into the CNS or when applied to dissociated
brain cells cultured in vitro indicates that cells on the brain
side of the barrier have insulin receptors and can respond to
insulin. The local application of insulin or glucose in the
caudate nucleus of rats increases the release of dopamine (Maccaleb
and Myers, 1979). Insulin applied electrophoretically to glucose
sensitive neurons in the lateral hypothalamus stimulates neuron
activity and facilitates neuronal discharge (Oomura, 1983).
Infusion of insulin into the cerebral ventricles of unanesthesized
baboons results in decreased food intake (Woods et al, 1979).

Cells cultured from rat fetal brain have a subpopulation of
neurons which bind to insulin and binding of insulin stimulates the
incorporation of thymidine and uracil into TCA-precipitable
products (Raizada et al, 1980; Weyhmeyer and Fellows, 1983).
Insulin also stimulates the activity of ornithine decarboxylase in
cultured rat brain cells (Yang et al, 1981) and chick brain cells
(Parker and Vernadakis, 1980). Sensory neurons can be cultured in
a serum-free medium if insulin and nerve growth factor are provided
(Bottenstein et al, 1980) and both insulin and an insulin-like
growth factor promote NGF-independent neurite formation by cultured
sensory neurons (Bothwell, 1982). Central neurons from 8-day
embryonic chick and newborn mouse can also be maintained if insulin
is present in the media (Skaper et al, 1982). These results
indicate that at least some neuronal cells are capable of
responding to insulin and that insulin or a related substance is
present on the brain side of the barrier.

There are a number of reports of insulin-like immunoreactivity
in brain of a variety of species (Dorn et al, 1980; Dorn et al,
1982; Dorn et al, 1981; Pansky and Hatfield, 1978). These studies,
although provocative, are not conclusive. A rigorous study which
deals with the possible artifacts and contradictory findings would
be very helpful. The current studies appear to indicate more
immunoreactive insulin in brain than is accounted for by
radioimmune assay (Yallow and Eng, 1983).

The most unambiguous demonstration of insulin in nerve cells is
in the vagal, sciatic and radial nerves of the cat. If the vagal
or sciatic nerves are ligated, immunoreactive insulin accumulates
on the cell body side of the ligature (Uvnäs-Moberg et al, 1982).

Electrical stimulation of sciatic nerve results in the release of
immunoreactive insulin (Uvnäs-Moberg and Uvnäs and Wallenstein,
1978). These investigators have not been able to visualize insulin
in cell bodies (Uvnäs-Wallenstein, (1981), but the presence of
amines within a cell may interfere with the interaction of peptides
and antibodies to them (Polak and Birchan, 1980). Since the
peripheral nervous system is not sequestered from the circulation,
the presence of insulin in peripheral nerve cells cannot be
conclusively interpreted to mean that insulin is synthesized there.

The best evidence that cells in the brain synthesize insulin is
that immunoreactive insulin can be detected in cultured brain cells
(Raizada, 1983; Birch et al, 1984; Weyhenmeyer and Fellows, 1983).
In one study, immunohistochemistry was used to identify a
subpopulation of neurons which stained with anti-insulin antibody
(Raizada, 1983). In this study, the cells were initially cultured
in media containing insulin, so the sequestration of insulin from
the medium can not be ruled out. The investigators did find that
the number of staining cells was reduced by 80% and the intensity
of staining decreased if the cultures were treated with
cyclohexamide for 24 hours; however, it is not clear if this
experiment was done in the presence or absence of serum, so the
effect of cyclohexamide could be on the synthesis of a factor
necessary for the stability of sequestered insulin or on the
synthesis of insulin itself. In the second study, fetal mouse
brain cells were cultured for 12 days in the absence of insulin or
serum and then assayed by RIA for insulin. These cells grew,
differentiated and also produced immunoreactive insulin (IRI; Birch
et al, 1984).

One of the most intriguing pieces of evidence that insulin can
be synthesized extrapancreatically is the persistence of insulin in
depancreatized chickens even though pancreatic polypeptide
decreases rapidly in these animals (Colca and Hazelwood, 1982).
The insulin detected cannot be accounted for by pancreatic remnants
since the animals were killed after 8 days and checked for
pancreatic remnants. There is also a report that insulin is
present in early chick embryos (dePablo et al, 1982). There are no
studies available that examine chicken brains for insulin-like
molecules.

Insulin receptors in the synaptosome fraction of olfactory
tubercle, hippocampus and hypothalamus were photoaffinity labeled
with a photoreactive analog of insulin. The receptors labeled by
this method were indistinguishable from each other, but differed in
size, antigenicity and carbohydrate composition from the insulin
receptor found on adipocytes (Heidenreich et al, 1983). This
difference can be interpreted in two ways: the insulin receptor
differs because it must function differently in the CNS from the
periphery or the receptor differs because it recognizes a different
protein.

The level of IRI reported varies over 4 logs not only between different groups, but even in reports from the same group (Baskin et al, 1983; Eng and Yallow, 1980; Eng and Yallow, 1981; Havrankova et al, 1978; Oomura and Kita, 1981. This difference has been widely discussed and debated, but not yet satisfactorily explained (for review see Havrankova et al, 1983; Yallow and Eng, 1983; Hendricks et al, 1983). Furthermore, there seems to be substantial variation both between species and between different individuals of the same species. The explanations for this variation include carelessness, differences in extraction procedures, inappropriate corrections for loss during preparations, differences in antibody titers or affinities and artifactual and irreproducible binding of the antibodies. One explanation recently advanced is that the material being detected by the anti-insulin antibodies is closely related, but not identical to insulin, and is different enough to have different stability and different properties in different purification schemes. A variation of this explanation is that the material is a mixture of authentic insulin and something closely related to insulin, or insulin associated with something which affects its reaction with anti-insulin antibodies as well as its other properties.

Other members of the insulin family, the insulin-like growth factors (IGF), may also play a role in the adult and developing CNS. The production of an IGF-like material by explants from rat pituitary and adult brain has been reported (Binoux et al, 1981). At least 2 forms of IGF-II have been identified in CSF (Haselbacher and Humbel, 1982) and receptors for both insulin and IGF are found in the developing brain and the adult brain of a variety of species from rats to man (Sara et al, 1982; 1983).

RECOMBINANT DNA TECHNOLOGY: A BRIEF DESCRIPTION

Recombinant DNA techniques have been used to isolate and amplify portions of eukaryotic DNA in bacterial cells. These studies have led to an explosion of information about the structure of eukaryotic genes including many specific to the CNS (Cold Spring Harbor Symposium, 1983). The information we have gained has profoundly changed some deeply held assumptions about the structure and regulation of eukaryotic genes. We once thought that the genetic information for a protein would have to be both contiguous and colinear; we have learned that many genes in higher cells are interrupted by long stretches of noncoding DNA. We once thought that the arrangement of DNA in somatic cells was the same as the arrangement of DNA in germline cells; we have found that DNA in the cells of the immune system rearranges to produce the diversity of immunoglobulins (for review see Lewin, 1983).

<u>Enzymes Used For Recombinant DNA Technology</u>

Recombinant DNA technology utilizes many of the enzymes which have been isolated and described over the years. DNA polymerase I is an enzyme from E. coli and was the first DNA synthesizing enzyme described (Kornberg, 1969). Ligase, also from E. coli, can heal the phosphodiester bond in a DNA chain with a break (Higgins and Cozzarelli, 1980). S1 nuclease is isolated from the mold aspergillus and degrades only single-stranded DNA or RNA (Ando, 1966; Britten et al, 1974). Reverse transcriptase is an enzyme found in retroviruses which can use RNA as a template for the synthesis of DNA (Baltimore, 1970; Temin and Mizutani, 1970; Green and Gerard, 1974). Terminal transferase was first isolated from calf thymus and adds triphosphates to the 3' end of a DNA molecule in a nontemplate dependent fashion (Bollum, 1974; Nelson and Brutlag, 1980). These enzymes, as well as others, were isolated and characterized over the last 20 years as part of basic research. Their use for the procedures described below was not anticipated or predicted, but those procedures would not be possible without them.

The restriction enzymes are among the most important enzymes for this work. They are endonucleases which recognize specific sites, called recognition sequences, in DNA and catalyze cleavage of the DNA at those sites (for review see Zabeau and Roberts, 1979). For the most part, restriction enzymes recognize either four or six base sequences within the DNA and cleave the two strands either directly opposite one another to leave flush ends or in a staggered fashion to leave "sticky" ends. For example, the enzyme Sma 1 and Xma 1 recognize the same sequence, but Sma 1 cleaves both strands in the middle to leave flush ends and Xma 1 cleaves in a staggered fashion. The result of digesting a DNA molecule with these two enzymes is illustrated below:

```
5'---CCCGGG----3'                5'---CCCGGG----3'
3'---GGGCCC----5'                3'---GGGCCC----5'

     ↓ digestion with Sma 1           ↓ digestion with Xma 1

----CCC + GGG----                ----C      + CCGGG----
----GGG   CCC----                ----GGGCC      C----
```

Over 50 restriction enzymes are available (Roberts, 1982). Each of these recognizes a different sequence and will therefore cleave a given piece of DNA into a characteristic set of fragments. This is important both for cloning DNA and for the analysis of DNA.

<u>Construction of Chromosomal DNA Libraries</u>

The genome of complex eukaryotic organisms such as primates, contains about 3×10^9 base pairs of DNA. Since single genes in

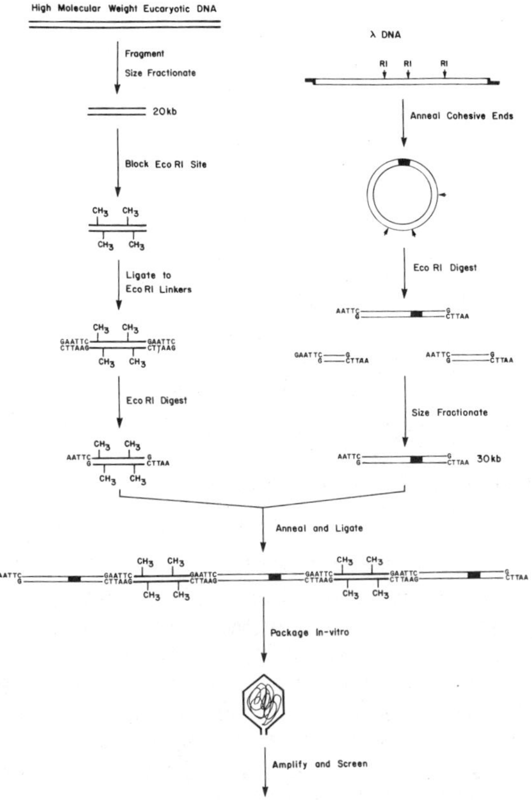

Figure 1: Construction of chromosomal DNA libraries.

such an organism represents only one millionth of the total DNA, it is impossible to study single genes using conventional techniques. However, any gene can, in principal, be isolated for study from a recombinant DNA library large enough to represent all of the sequences present in the genome. The feasibility of constructing such libraries was first demonstrated using organisms with small genomes (for example Wensink et al, 1974; Clark and Carbon, 1976). Technical advances made it possible to construct similar libraries from more complex organisms such as silkmoth, rabbit and human being (Maniatis et al, 1978; Lawn et al, 1978).

The strategy used to generate such libraries is illustrated in figure 1. High molecular weight DNA from the organism is broken into fragments either by physically shearing the DNA and treating it with S1 nuclease, or by partially digesting the DNA with a

restriction enzyme which leaves blunt ends. If the latter method
is used, a restriction enzyme which cleaves DNA very frequently is
used in order to approximate a random shear. Generally two enzymes
are used so that a cleavage site can be expected to occur about
every 200 bases. After fragmentation, DNA fragments of about
20,000 base pairs are selected by centrifugation through a sucrose
gradient and treated with Eco Rl methylase, an enzyme which
transfers a methyl group from S-adenosyl-L-methione to one of the
adenines in the Rl restriction site. When this adenine is
methylated, the site can no longer be cleaved by the enzyme Eco Rl
(Dugaiczyk et al, 1974; Greene et al, 1979). Small pieces of DNA
have been chemically synthesized which contain an Eco Rl
recognition site (Itakura and Riggs, 1980). These linkers are
covalently attached to the ends of the eukaryotic DNA fragments
using the enzyme DNA ligase. The DNA is then digested with the
restriction enzyme Eco Rl to generate cohesive ends. The
eukaryotic DNA must then be inserted into a vector DNA which can
replicate in bacterial cells.

The vector most often used for this type of cloning is a
modified lambda phage (Blattner et al, 1977; Leder et al, 1977).
These phages contain restriction enzyme sites which define internal
fragments, do not encode any functions needed for virus replica-
tion, and so can be replaced by any DNA of appropriate length. A
variety of phage vectors have been constructed to allow the cloning
of DNA fragments of different lengths. The ends of lambda are
complementary to each other, so the vector is circularized, ligated
and digested with Eco Rl. The small internal fragments are
separated from the connected end fragments by centrifugation
through sucrose.

The vector DNA is then mixed with the eukaryotic DNA and the Rl
ends are allowed to anneal under conditions which optimize the
formation of concatemers. The concatemers are ligated and the DNA
is packaged into phage particles using an in vitro packaging system
(Enquist and Sternberg, 1980). In vitro packaging involves the
preparation of extracts from two different sets of bacterial cells
carrying defective lambda genomes in the chromosome of the
bacteria. The lambda genomes have been mutated so that they cannot
excise from the bacterial chromosome and so that each one lacks a
different function necessary for the assembly of complete phage
particles. As a result of these mutations, a mixture of the two
extracts contains all the functions necessary for packaging except
DNA. When the recombinant DNA is added to a mixture of the two
extracts, phage particles are assembled which contain a piece of
DNA which is about 40,000 base pairs long and which has the
cohesive ends of the phage DNA at each end. The phage particles
are then used to infect bacterial cells. Using this procedure one
can easily obtain a library of recombinant phages which contains
all of the sequences present in the genome under study.

Libraries such as the one described above are constructed so as
to study individual genes. In order to study a given gene, one
must be able to identify it. In most cases, the gene of interest
is identified by hybridization to radiolabled DNA containing a
portion of the sequence being sought. Such a probe can be
generated as described below.

<u>Construction of double-stranded cDNA libraries:</u>

The collection of bacterial colonies which contains DNA copies
of the RNAs present in the starting material is called a
double-stranded copy DNA (ds-cDNA) library. Many such libraries
have been constructed; for example, we have constructed libraries
from the RNA of several human fetal tissues (Kurnit et al, 1982).
Recombinant DNA libraries of this type are constructed because
sequences encoding specific proteins are found in the form of mRNA
in cells where that protein is synthesized. Insulin, for example,
is made in the B-cells of the pancreas; those cells contain insulin
mRNA and other cells do not. To isolate the sequence for insulin,
it must be converted from RNA into DNA, which can be cloned (see
Efstratiadis and Villa-Komaroff, 1979 for review). This process is
illustrated in Figure 2. First the poly(A) containing RNA is
isolated from the cells; this fraction contains most of the
messenger RNA in the cell. Reverse transcriptase is then used to
copy the RNA into a complementary strand of DNA. This enzyme, like
all DNA synthesizing enzymes, requires a primer, so a small piece
of oligo(dT) is annealed to the poly(A) at the 3' end of the
template RNA. After synthesis, the RNA template is destroyed and a
second strand of DNA is synthesized using either reverse
transcriptase or DNA polymerase. The second strand of DNA is
covalently attached to the first strand because the enzyme utilizes
the 3' end of the first strand as a primer to begin synthesis of
the second strand. The covalent linkage can be broken by treating
the doubled-stranded product with S1 nuclease.

The DNA must now be attached to a vector. In this example, a
plasmid is used instead of phage DNA. A plasmid is a small
circular double-stranded DNA which is capable of replicating
autonomously in a bacterial cell. The plasmids used for
recombinant DNA work also carry a gene or genes specifying
resistance to antibiotics. The plasmid most often used is pBR322
(Bolivar et al, 1977). This plasmid is small (4300 base pairs),
carries resistence to both ampicillin and tetracycline and the
nucleotide sequence of the entire plasmid has been determined
(Sucliff, 1978). There are several restriction enzymes whose
recognition sequence occurs only once on the plasmid.

To attach the ds-cDNA to the plasmid, the enzyme terminal
transferase is often used to extend the 3' ends of the DNA with a
homopolymer and then to anneal the DNA to linearized plasmid to

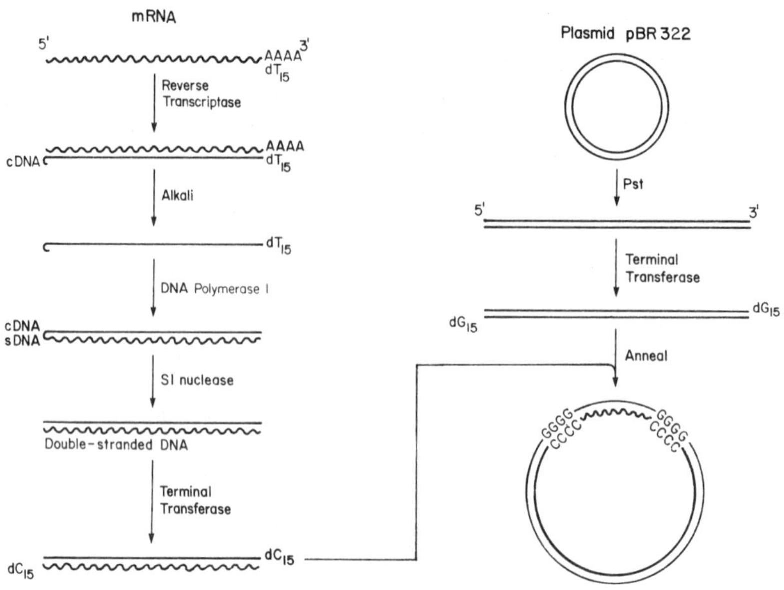

Figure 2: Construction of ds-cDNA libraries.

which complementary homopolymers have been added (Nelson and
Brutlag, 1980; Lobban and Kaiser, 1973). There are two
combinations which can be used: poly(dA)-poly(dT) or
oligo(dG)-oligo(dC). The latter combination has two advantages:
the homopolymer tracts can be short (on the order of 10 bases) and
the use of dG and dC allows the reconstruction of a recognition
site for the enzyme Pst 1 at each end of the inserted DNA, as
illustrated in figure 3. However, the dG-dC tails have some
disadvantages: They are extremely stable and it can be difficult to
separate the strands of DNA fragments with these ends. The tails
will anneal to G-C rich regions of DNA when hybidizations are done
under nonstringent criteria. Because of these problems, it is
sometimes advantageous to use dA-dT joining. These homopolymer
tails must be much longer, on the order of 100 to 200 bases and one
cannot reconstitute a convenient restriction site for excision of
the insert. However, methods have been developed to excise inserts
cloned using poly(dA)-(dT) joining (Hofstetter et al, 1976; Goff
and Berg, 1978).

The ds-cDNA can also be attached to the plasmid by attaching
synthetic linkers to the ds-cDNA. The efficient use of linkers
makes it possible to prepare cDNA libraries in the lambda phage
vectors (Young and Davis, 1983). These vectors are particularly
attractive because they may allow the isolation of many proteins

74

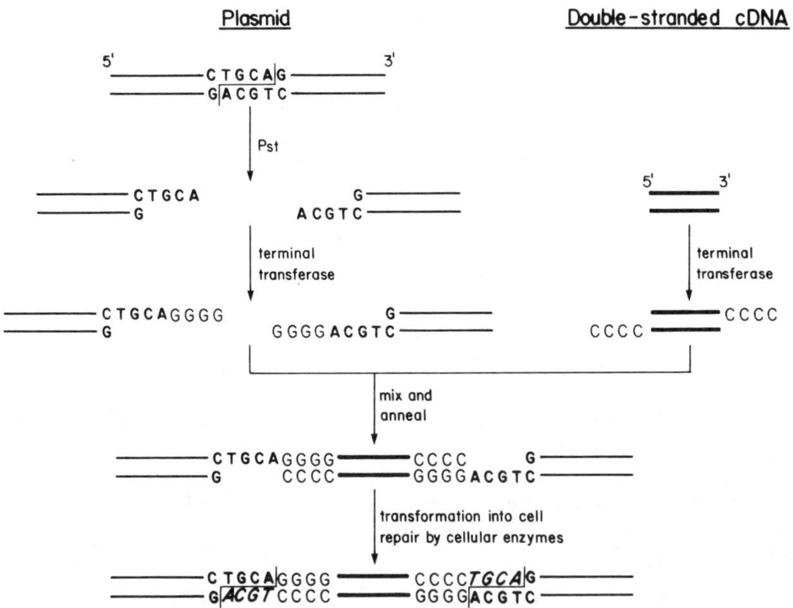

Figure 3: Reconstruction of the Pst-1 recognition site.

which have been identified by specific antibodies but whose amino
acid sequence is not known. The use of phage vectors also allows
the construction and easy handling of large numbers of
recombinants. This is particularly important in the CNS where 35
to 45% of the genome is transcribed into RNA (Van Ness et al, 1979,
Chikaraishi, 1979; Chaudhari and Hahn, 1983).

After the ds-cDNA has been attached to the plasmid, the DNA
must be put into a bacterial host and those bacteria receiving the
recombinant molecules must be identified. Ordinarily, DNA cannot
pass through the cell wall of bacterial cells, but E. coli cells
can be rendered competant to take up DNA by treatment with calcium
(Mandel and Higa, 1970). The cells which take up the DNA can be
selected by virtue of the drug resistence encoded by the plasmid.
The unique site for the restriction enzyme Pst 1, which is
frequently used for cloning, lies in the middle of the plasmid gene
encoding ampicillin resistence, so cells which received a
recombinant DNA molecule can be identified by virtue of their
resistence to tetracycline and their sensitivity to ampicillin.
There are other unique restriction sites in pBR322 which lie in the
gene encoding tetracycline resisence which can be used in an
analogous fashion.

Once the colonies containing inserts are identified, the
sequence of interest must be identified. The most straightforward

method to identify clones containing the desired sequence is to
hybridize RNA (or cDNA) to DNA in the bacterial colonies. In some
cases, a related sequence can be used in the hybridization assay.
For example, the human insulin sequence was isolated from a cDNA
library by virtue of its homology to the already cloned rat insulin
sequence (Bell et al, 1979; Sures et al, 1980, Song and
Villa-Komaroff, 1984). In most cases, the RNA of interest will be
a minor fraction of the total RNA and will have no unique features
which allow its physical purification. In such cases, the sequence
of interest must be identified by less direct methods.

One useful method depends on the ability to identify the
protein encoded by the sequence of interest. Such identification
can be the position of the protein on a gel, the recognition of the
protein by antibodies, or, in some cases, the biological activity
of the protein. This method utilizes the ability to translate mRNA
in vitro. One must first demonstrate that translation of RNA from
the starting tissue results in the synthesis of the protein of
interest. For example, the translation of RNA from rat insulinoma
results in the synthesis of a protein which is precipitable with
anti-insulin antibody (Lomedico and Sanuders, 1976). The
translation of RNA from induced white blood cells in frog oocytes
results in the production of interferon, as assayed by the ability
of the products of such a translation to protect cells from virus
infection (Nagata et al, 1980).

Large amounts of DNA from the cells containing the recombinant
molecules are then prepared. (If the library is very large, one
can pool DNA from many clones.) The DNA is rendered
single-stranded and covalently attached to a filter. RNA from the
starting material is then passed through the filter under
conditions where homologous sequences can anneal. Thus any RNA
complementary to the DNA on the filter will be retained on the
filter and the other RNA will pass through. The RNA can then be
eluted from the filter, translated in vitro, and the products
analyzed for the synthesis of the protein of interest.

Using the clones isolated

Once a clone containing the desired sequence has been
identified, the sequence can be grown in large amounts for further
study. The nucleotide sequence of the RNA and the amino acid
sequence of the protein encoded by the RNA can be easily determined
by sequencing the cloned DNA (Maxam and Gilbert 1977; Sanger et al,
1977). The ability to rapidly determine the nucleotide sequence of
a DNA molecule has made possible the detailed analysis of the
structure of many proteins and genes.

The sequence can also be manipulated so that the eukaryotic
coding sequences are under the control of bacterial sequences
necessary for transcription and translation. This will result in

the production of the eukaryotic protein in the bacterial cells.
Hormones, enzymes and viral surface peptides have all been produced
in bacteria (for review, see Villa-Komaroff, 1980; Gilbert and
Villa-Komaroff, 1980). Such proteins appear to be biologically
active.

The ds-cDNA clone can also be used to screen a chromosomal
library for the gene encoding it. The chromosomal clones so
isolated can be used to elucidate the structure and arrangement of
the eukaryotic genes.

IDENTIFICATION OF NOVEL INSULIN-LIKE SEQUENCES IN BRAIN RNA

We have used nucleic acid hybridization to determine whether we
could identify RNAs encoding insulin or sequences related to
insulin in RNA from brain. In one preliminary report (Giddings et
al, 1981), no insulin-like RNA was detected in RNA from various
regions of the brain. However, using cloned sequences encoding rat
and human proinsulin, we detect sequences homologous, but not
identical, to insulin in RNA from adult rat, fetal mouse and fetal
human brain. Some of these sequences have more homology to insulin
then do the sequences encoding IGF-II (Janset et al, 1983). These
experiments are done by isolating polyadenylated RNA from the
tissues by standard procedures (Chirgwin et al; Aviv and Leder,
1972), separating the RNA species of different sizes by
electrophoresis on denaturing gels (Lehrach et al, 1977),
transferring the RNA to nitrocellulose or nylon filters (Thomas,
1980) and hybridizing to radiolabeled DNA encoding the sequence of
interest under different conditions. Conditions of hybridization
where nucleic acids which are only partly complementary form stable
hybrids are called nonstringent. As the stringency of the
conditions are increased (by decreasing salt, increasing
temperature or increasing formamide concentration) the hybrids must
be increasingly well matched in order to form a stable hybrid.

Figure 4A shows hybridization of ^{32}P-labeled cloned cDNA
encoding rat proinsulin (Villa-Komaroff et al, 1978) to RNA from
rat pancreas and adult rat brain. There is a species of RNA in the
adult rat brain which co-migrates with the pancreatic insulin
mRNA. This RNA does not correspond to the pancreatic insulin mRNA,
however, because the hybrid formed between the rat brain RNA and
the cloned insulin sequence is not as stable as the hybrid formed
between the pancreatic insulin RNA and the same radiolabeled DNA
(Figure 4B). When we examine RNA from fetal, neonatal and adult
mouse brains, we detect sequences homologous to insulin in fetal
brain RNA, but not in RNA from neonatal brain or adult brain (Song
and Villa-Komaroff, unpublished observations). In this case, the
RNA is not the same size as the pancreatic insulin RNA, but is
larger. This RNA also cannot be maintained in a stable hybrid
under conditions where the pancreatic insulin RNA can form a stable
hybrid.

 A 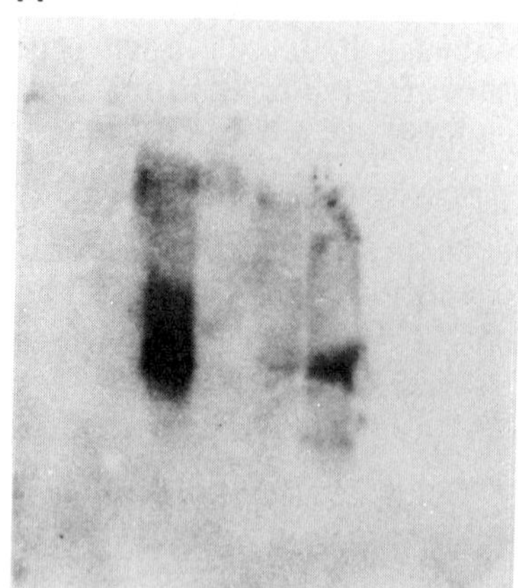 B 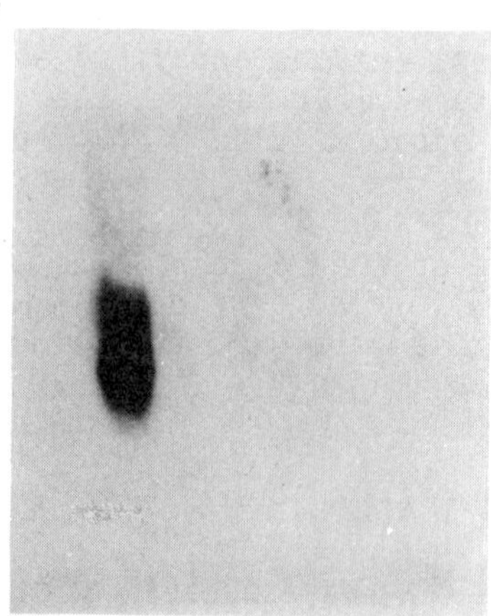

Figure 4:   Hybridization of cDNA encoding rat insulin to RNA from
             rat pancreas and rat brain.   A. ^{32}P-labeled DNA
             encoding rat insulin was hybridized to RNA from rat
             pancreas (left lane) and rat brain (right lane) in 10%
             formamide, 0.6M NaCl, 37^0 and washed in 50% formamide
             0.6M NaCl at 37^0 and exposed to x-ray film for 18
             hours.   B.   The filter described in A was washed in 50%
             formamide, 52^{0}C, 0.6M NaCl and exposed to x-ray film
             for 24 hours.

     When RNA from human fetal brains is hybridized to DNA encoding
rat insulin, at least 4 RNAs can be visualized (Figure 5).   These
RNAs are much larger than the pancreatic mRNA, and also appear not
to correspond to pancreatic insulin RNA.   Several RNAs from the
human fetal brain hybridize to the rat insulin probe at low
stringency (Fig. 5, Lane 3).   The most prominent RNA species
visualized are 2100, 2700, 2900, and 3500 bases in size, but there
are also RNA species of 3900, 6200 and 7800 bases.   At moderate
stringency the 3500, 3900, 6200 and the 7800 base RNAs are still
apparent.   An RNA the same size as 18S continues to hybridize at
moderate stringency and so may represent a moderately homologous
sequence (Fig. 5, Lane 4).   At high stringency, the 3500, the 3900
and the 7800 bands are still visible (Figure 5, Lane 5).   The 3500
and 3900 base RNAs appear to be the most closely related to insulin
since the relative intensity of the signal does not change after
the high stringency wash.   Essentially the same pattern is seen
when the human insulin sequence is used as the radiolabeled DNA
(Gonzalez et al, in preparation).

     Our results suggest that there are RNA sequences related, but
not identical to insulin, in the CNS.   These RNAs presumably encode
proteins which are related, but not identical, to insulin.   The

78

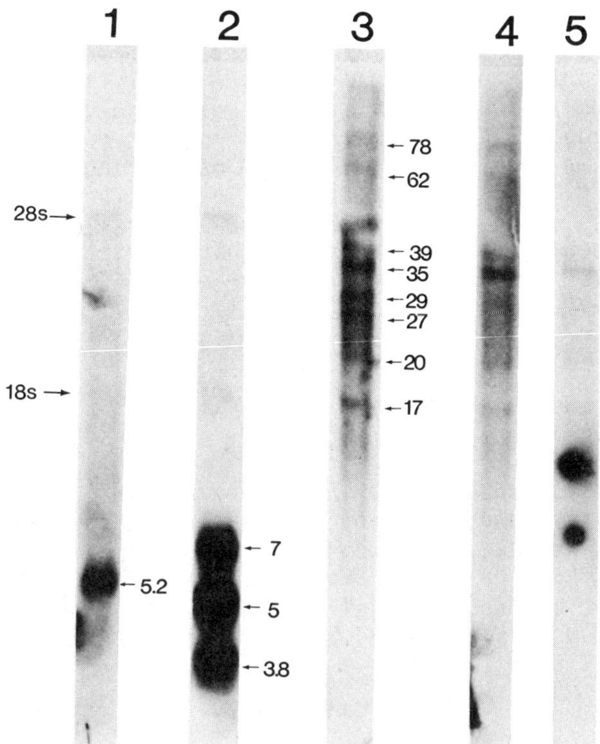

Figure 5: Hybridization of cDNA encoding rat insulin to RNA from rat pancreas, human insulinoma and human fetal brain. ^{32}P-labeled DNA encoding rat insulin was hybridized to RNA from rat pancreas (Lane 1); human insulinoma (Lane 2) and human fetal brain (Lane 3) at 25% formamide, 0.75M NaCl, 42°. Lane 4 is the same as Lane 3 washed at 35% formamide; Lane 5 is the same as Lane 3 washed at 50% formamide.

interaction of such peptides with different anti-insulin antibody preparations may explain the wide variation in results obtained by different groups. The information we have is clearly incomplete. We have not examined the RNA of rat fetal brains. We do not know if adult mice brains have no insulin related RNAs or simply have much less than do rats. We do not know where these sequences are located - such a localization experiment is best done with a cloned sequence. To isolate such sequences, we enriched for insulin-related sequences by hybridizing RNA from fetal mouse brains to DNA encoding human insulin covalently attached to a cellulose support (Goldberg et al, 1979). The RNA which hybridized was eluted from the cellulose, and used as template for the synthesis of ds-cDNA which was inserted into the plasmid vector

pUC8 by the d(G)-d(C) homopolymer addition procedure described
above. The nucleic acid sequence of two of the clones we have
examined indicates that the inserted DNA consists of a small
segment of human sequence attached to murine sequence. These
clones are probably the result of small pieces of DNA from the
affinity column hybridizing to the RNA and acting as a primer in
the generation of the cDNA (Song et al, unpublished results). We
are using these clones to screen new cDNA libraries made from mouse
brain RNA. We are in the process of characterizing cDNAs encoding
the mouse fetal brain sequences and expect that the sequences of
these RNAs will help to clarify the role of insulin-like sequences
in the CNS.

ACKNOWLEDGEMENTS

This work was supported by NIH grant no. GM26028; and a Basil
O'Conner Starter Research Grant no. 5-344 from the March of Dimes
Birth Defects Foundation. S.H.Y. was supported by a grant from the
Ministry of Education of the People's Republic of China, A.G. was
supported by a postdoctoral fellowship from the Scientific Council
of UMMC and P.D. was supported by a postdoctoral fellowship from
the American Cancer Society.

REFERENCES

Ando T., 1966, A nuclease specific for heat-denatured DNA is
    isolated from a product of Aspergillus oryzae. Biochem.
    Biophys. Acta 114:158-68
Aviv H. and Leder P., 1972, Purification of biologically active
    globin messenger RNA by chromatography on oligothymidylic
    acid-cellulose, Proc. Natl. Acad. Sci. USA 69: 1408-1412.
Baltimore D., 1970, RNA-dependent DNA polymerase in virions of RNA
    tumor viruses. Nature 226:1209-1211.
Baskin D.G., Porte D., Guest K. and Dorsa D.M., 1983, Regional
    concentrations of insulin in the rat brain, Endocrinology 112:
    898-903.
Bell G. I., Swain W.F., Pictet R., Cordell B., Goodman H.M. and
    Rutter W.J., 1979, Nucleotide sequence of a cDNA clone encoding
    human preproinsulin. Nature 282:525-527.
Berggren V., Engel J. and Liljequist S., 1983, Differential effects
    of insulin on brain monoamine metabolism in rats, Acta
    Phrmacol. et toxicol 53: 39-43.
Birch N.P., Christie D.L. and Renwick A.G.C., 1984, Immunoreactive
    insulin from mouse brain cells in culture and whole rat brain,
    Biochem. J. 217: in press.
Binoux M., Hossenlopp P., Lassarre C. and Hardouin N., 1981,
    Production of insulin-like growth factors and their carrier by

rat pituitary gland and brain explants in culture, FEBS Letts.
24: 178-184.

Blattner F.R., Williams B.G., Blechl A.E., Thompson K.D.,
Faber H.E., Furlong L.A., Grunwald D.J., Kiefer D.O., Moore
D.D., Schumm J.W., Sheldon E.L., and Smithies O., 1977, Charon
phages: safer derivatives of bacteriophage lambda for DNA
cloning. Science 196: 161-169.

Bolivar F., Rodriguez R.L., Greene P.J., Betlach M.C.,
Heyneker H.L., Boyer H.W., Crossa J.H. and Falkow S., 1977,
Construction and characterization of new cloning vehicles II. A
multipurpose cloning system. Gene 1:95-113.

Bollum R., 1974, Terminal deoxynucleotide transferase, The
Enzymes 10: 195.

Bothwell M., 1982, Insulin and somatomedin MSA promote nerve growth
factor in dependent neurite formation by cultured chick dorsal
root ganglionic sensory neurons, J. Neurosci. Res. 8: 225-231.

Bottenstein J.E., Skaper S.D., Varon S.S., Sato G.H., 1980,
Selective survival of neurons from chick embryo sensory
ganglionic dissociates utilizing serum-free supplemented
medium. Exp. Cell Res. 125: 183-190.

Britten R.J., Graham D.E. and Neufeld B.R., 1974, Analysis of
repeating DNA sequences by reassociation, in Methods in
Enzymology ed. L. Grossman and K. Moldare, Academic, NY. pp.
363-418.

Chaudhari N. and Hahn W.E., 1983, Genetic expression in the
developing brain, Science 220: 924-928

Chikaraishi D.M., 1979, Complexity of cytoplasmic polyadenylated
and nonpolyadenylated rat brain ribonucleic acids,
Biochemistry 18: 3249-3256.

Chirgwin J.M., Przybyla A.E. and Rutter W.J., 1979, Isolation of
RNA in the presence of ribonuclease with quanidinium
thiocyanate, Biochemistry 18: 5294-5298.

Clarke L. and Carbon J., 1976, A colony bank containing synthetic
Col El hybrid plasmids representative of the entire E. coli
genome. Cell 9:91-99

Colca J.R. and Hazelwood R.L., 1982, Persistence of
immunoreactive insulin, glucagon and pancreatic polypeptide in
the plasma of depancreatized chickens, J. Endocrin. 92: 317-326.

Cold Spring Harbor Symposia on Quantitative Biology, 1983,
Molecular Neurobiology vol. 48, part 1 and 2.

DeMontis M.G, Olianas M.C., Haber B. and Tagliamonte A., 1978,
Increase in large neutral amino acid transport into brain by
insulin. J. Neurochem. 30: 121-124.

dePablo F., Roth J., Hernandez E. and Pruss R.M., 1982, Insulin is
present in chicken eggs and early chick embryos, Embryology
111: 1909-1916.

Dorn A., Bernstein H-G, Hahn H-J, Ziegler M. and Rummelfänger H.,
1981, Insulin immunoreactivity of rodent CNS: apparent species
differences but good correlation with raioimmunological data,
Histochem. 71: 609-616.

Dorn A., Bernstein H-G, Kostmann G., Hahn H-J, and Ziegler M., 1980, An immunofluorescent reaction appears to insulin antiserum in different CNS regions of two rat species, Acta Histochem. 66: 276-278.

Dorn A., Bernstein H-G, Rinne A, Hahn H-J and Ziegler M., 1982, Insulin-like immunoreactivity in human brain: a preliminary report, Histochem. 74: 293-300.

Dugaiczyk A., Hedgpeth J., Boyer H.W. and Goodman H.M., 1974, Physical identity of the SV40 deoxyribonucleic acid sequence recognized by the Eco R1 restriction endonuclease and modification methylase. Biochemistry 13:503-512.

Efstratiadis A. and Villa-Komaroff L., 1979, Cloning of double-stranded cDNA in Genetic engineering: Principles and methods vol 1, edited by Hollaender, A., and Setlow, J. New York Plenum pp 15-36.

Eng J. and Yallow R.S., 1980, Insulin recoverable from tissues, Diabetes 29: 105-109.

Eng J. and Yallow R.S., 1981, Evidence against extrapancreatic insulin synthesis, Proc. Natl. Acad. Sci. USA 78: 4576-4578.

Enquist L. and Sternberg N., 1980, In vitro packaging of DAM vectors and their use in cloning DNA fragments, in Methods in Enzymology, v.68, ed. R. Wm, Academic Press, NY pp. 281-298.

Frank J.L.H. and Pardridge W.M. Insulin binding to brain microvessels in Advances in Metabolic disorders 10: CNS regulation of carbohydrate metabolism" A.J. Szalo, e., Academic New York, 1983.

Giddings S., Chirgwin J. and Permutt M.A., 1981, Expression of the insulin gene in pancreatic and extra-pancreatic tissues, Endocrinology Suppl. A-624, 238.

Gilbert W. and Villa-Komaroff L., 1980, Useful proteins from recombinant bacteria. Scientific American 242-4: 74-94.

Goff S.P. and Berg P., 1978, Excision of DNA segments introduced into cloning vectors by the poly(dAT, method. Proc. Nat'l. Acac. Sci. USA 76:106-110.

Goldberg M.L., Lifton R.P., Starke G.R. and Williams J.G., 1979, Isolation of specific RNAs using DNA covalently linked to diazobenzlomethyl cellulose or paper, in Methods in Enzymology 68: 206-222 ed. R. Wu, Academic Press.

Gonzalez A., Wentworth B., Dobner P. and Villa-Komaroff L., RNA from nonpancreatic tissues contains sequences homologous to insulin, in preparation.

Goodner C.J. and Berrie M.A., 1977, The failure of rat hypothalamic tissues to take up labeled insulin in vivo or to respond to insulin in vitro, Endocrinology 101: 605-612.

Goodner C.J., Horn F.G. and Berrie M.A., 1980, Investigation of the effect of insulin upon regional brain glucose metabolism in the rat in vivo, Endocrinology 107: 1827-1832.

Green M. and Gerard G.F., 1974, RNA-directed DNA polymerase - properties and functions in oncogenic RNA viruses and cells, in Progress in Nucleic Acid Res. Mol. Biol., ed. W.E. Cohn, vol. 14, Academic Press, NY p. 187-334.

Greene P.J., Poonian M.S., Nussbaum A.L., Tobias L., Garfin D.E., Boyer H.W. and Goodman H.M., 1975, Restriction and modification of a self-complementary octanucleotide containing the Eco Rl substrate J. Mol. Biol. 99:237-261.

Haselbacher G. and Humbel R., 1982, Evidence for two species of insulin-like growth factor II (IGF-II and "Big" IGFII) in human spinal fluid, Endocrinology 110: 1822-1824.

Havrankova J. and Roth J., 1978, Insulin receptors are widely distributed in the central nervous system of the rat, Nature 272: 827-829.

Havrankova J., Roth J. and Brownstein M.J., 1979, Concentrations of insulin and of insulin receptors in the brain are independent of peripheral insulin levels, J. Clin. Invest. 64: 636-642.

Havrankova J., Schmechel D., Roth J. and Brownstein M., 1978, Identification of insulin in rat brain, Proc. Natl. Acad. Sci. USA 75: 5737-5741.

Heidenreich K.A., Zahniser N.R., Berhanu P., Brandenburg D. and Olefsky J.M., 1983, Structural differences between insulin receptors in the brain and peripheral target tissues, J. Biol. Chem. 258: 8527-8530.

Hendricks S.A., Roth J., Rishi S. and Becker K.L., 1983, Insulin in the nervous system, in Brain Peptides ed D.T. Krieger, M.J. Brownstein and J.B. Martin, John Wiley and Sons.

Hertz M.M., Paulson O.B., Barry D.I., Christiansen J.S. and Svendsen P.A., 1981, Insulin increases glucose transfer across the blood-brain barrier in man, J. Clin. Invest. 67: 597-604.

Higgins N.P. and Cozzarelli N.R., 1980, DNA joining enzymes: A review, in Methods in Enzymology v.68 ed. R. Wu, Academic Press, New York, pp. 50-71.

Hofstetter H., Schambock A., Van den Berg, J. and Weisman, C., 1976, Specific Excision of the Inserted DNA Segment from Hybrid Plasmids Constructed by the poly(dA) poly(dT) method. Biochem. Biophys. Acta 454: 587-591.

Itakura K. and Riggs A.D., 1980, Chemical DNA synthesis and recombinant DNA studies. Science 209:1401-1405.

Jansen M., van Schaik F.M.A., Ricker A.T., Bullock B., Woods D.E., Gabbay K.H., Nussbaum A.L., Sussenback J.S. and Van de Brande J.L., 1983, Sequence of cDNA encoding human insulin-like growth factor 1 precursor, Nature 306: 609-611.

Kornberg A., 1969, Active center of DNA polymerase. Science 163: 1410-1418.

Kurnit D.M., Wentworth B.M., DeLong L. and Villa-Komaroff L., 1982, The construction of cloned libraries from mRNA of human fetal tissues, Cytogen. Cell Gen. 34: 193-203.

Lawn R.M., Fritsch E.F., Parker R.C., Blake G., and Maniatis, 1978, The isolation and characterization of linked α-and β-globin genes from a cloned library of human DNA. Cell 15:1157-1174, 1978.

Leder P., Tiemeier D. and Enquist L., 1977, EK2 derivatives of bacteriophage lambda useful in the cloning of DNA from higher organisms: the λgtWES, system. Science 196: 175-177.

Lehrach H., Diamond D., Wozney J.M. and Boedtker, 1977, RNA
    molecules weight determination by gel electrophoresis under
    denaturing conditions, Biochemistry 16: 4743-4751.
LeRoith D., Hendrick S.A., Lesniak M.A., Rishi S., Becker K.I.,
    Havrankova J., Rosenzwieg J.L., Brownstein M.J. and Roth J.,
    1983, Insulin in brain and other extrapancreatic tissues of
    vertebrates and nonvertebrates, Advances in Metabolic Disorders
    10: 303-339.
Lewin B., 1983, Genes. John Wiley and Sons, Inc. NY.
Lobban P.E., and Kaiser A.D., 1973, Enzymatic end-to-end joining
    of DNA molecules. J. Mol. Biol. 78:453-471
Lomedico P.T., Saunders G.F., 1976, Preparation of pancreatic mRNA:
    cell-free translation of an insulin-immunoreactive
    polypeptide. Nucleic Acids Res. 3:381-391.
Lorenzi M., Karam J.H., McIlroy M.B. and Forsham P., 1980,
    Increased growth hormone response to dopamine infusion in
    insulin-dependent diabetic subjects. J. Clin. Invest. 65:
    146-153.
Maccaleb M.L. and Meyers R.D., 1979, Striatal dopamine release is
    altered by glucose and insulin during push-pull perfusion of
    the rat caudate nucleus, Brain Res. Bull. 4: 651-656.
Mackenzie R.G. and Trulson M.E., 1978, Effects of insulin and
    streptozotocin-induced diabetes on brain tryptophan and
    serotonin metabolism in rats, J. Neurochem. 30: 1205-1208.
Mandel M. and Higa A., 1970, Calcium-dependent bacteriophage DNA
    infection. J. Mol Biol. 53:159.
Maniatis T., Hardison R.C., Lacey E., Lauer J. O'Connell C.,
    Quon D., Gek Kee Sim and Efstratiadis A., 1978, The isolation
    of structural genes from libraries of eukaryotic DNA. Cell
    15:687-701.
Margolis R.V. and Altszuler N., 1967, Insulin in the cerebrospinal
    fluid, Nature 215: 1375-1376.
Maxam A. and Gilbert W., 1977, A new method for sequencing DNA.
    Proc. Natl. Acad. Sci. USA 74:560-564
Nagata S., Taira J.J., Hall A., Johnsrud L., Streuli M.,
    Ecsodi J., Boll W., Cantell K., and Weisman C., 1980, Synthesis
    in E. coli of a polypeptide with human leukocyte interferon
    activity. Nature 284:316-320.
Nelson T. and Brutlag D., 1980, Addition of homopolymers to the
    3'-ends of duplex DNA with terminal transferase, in Methods in
    Enzymology, v. 68, ed. R. Wu, Academic Press, NY, pp.41-50.
Oomura Y. and Kita H., 1981, Insulin acting as a modulator of
    feeding through the hypothalamus, Diabetologia 20: 290-298.
Oomura Y., Glucose and neuronal activity, Advances in metabolic
    disorders 10: CNS regulation of carbohydrate metabolism, A.J.
    Szalo, ed., Academic, New York, 1983).
Pacold S.T. and Blackard W.G., 1979, Central nervous system insulin
    receptors in normal and diabetic rats, Endocrinology 105:
    1452-1457.
Pansky B. and Hatfield J.S., 1978, Cerebral localization of insulin
    by immunofluorescence, Amer. J. Anat. 153: 459-467.

Parker K. and Veradakis A., 1980, Stimulation of ornithine
    decarboxylase activity in neural cell culture: potential role
    of insulin, J. Neurochem. 35: 155-163.
Polak J.M. and Bruchan A.M.J., 1980, Motilin-cellular origin in
    human gut, J. Histochem., Cytochem. 28: 617 (abs).
Raizada M.K., 1983, Localization of insulin-like immunoreactivity
    in the neurons from primary cultures of rat brain, Exp. Cell
    Res. 14: 351-357.
Raizada M.K., Yang J.W. and Fellows R.E., 1980, Binding of [^{125}I]
    insulin to specific receptors and stimulation of nucleotide
    incorporation in cells cultured from rat brain, Brain Res. 200:
    389-400.
Ratzmann K.P. and Hampel R., 1980, Glucose and insulin
    concentration patterns in cerebrospinal fluid following
    intravenous glucose injection in humans, Endokrinologie, 76:
    185-188.
Roberts R.J., 1980, Directory of restriction endonucleases, in
    Methods in Enzymology v. 68, ed. R. Wu, Academic Press, NY, pp.
    27-40.
Sanger F., Nicklen S. and Coulson A.R., 1977, DNA sequencing with
    chain-terminating inhibitors. Proc. Natl. Acad. Sci. USA
    74:5463-5467.
Sara V.R., Hall K., Misaki M., Fryklund L., Christensen N. and
    Wetterberg L., 1983, Ontogenesis of somatomedin and insulin
    receptors in the human fetus, J. Clin. Inves. 71: 1084-1094.
Sara V.R., Hall K., Holtz H.V., Humbel R., Sjögren B. and
    Wetterberg L., 1982, Evidence for the presence of specific
    receptors for insulin-like growth factors 1 (IGF-1) and 2
    (IGF-2) and insulin throughout the adult human brain, Neurosci.
    Letts. 34: 39-44.
Skaper S.D., Selak I. and Varon S., 1982, Molecular requirements
    for survival of cultured avian and rodent dorsal root
    ganglionic neurons responding to different trophic factors, J.
    Neurosci. Res. 8: 251-261.
Song H-Y and Villa-Komaroff L., 1984, Isolation and
    characterization of a double-stranded cDNA clone encoding
    preproinsulin, Scientia Sinica, in press.
Sures I., Goeddel D.V., Gray A. and Ullrich A., 1980, Nucleotide
    sequence of human preproinsulin complementary DNA. Science
    208:57-59.
Sutcliffe J.G., 1978, The complete nucleotide sequence of the
    Escherichia coli plasmid pBR322. Cold Spring Harbor Symp.
    43:77-90.
Temin H. and Mizutani S., 1970, RNA-dependent DNA polymerase in
    virons of Rous sarcoma virus. Nature 226:1211-1213.
Thomas P.S., 1980, Hybridization of denatured RNA and small DNA
    fragments transferred to nitrocellulose, Proc. Natl. Acad.
    Sci. USA 77: 5201-5205.
Uvnäs B. and Uvnäs-Wallenstein K., 1978, Insulinergic nerves to the
    skeletal muscles of the cat? Acta Physiol. Scand. 103: 346-348.

Uvnäs-Moberg K., Posloncec B., Hagerman M., Castensson S., Rubio C. and Uvnäs B., 1982, Occurence of an insulin-like peptide in extracts of peripheral nerves of the cat and in extracts of human vagal nerves. Acta Physiol. Scand. 115: 471-477.

Uvnäs-Wallensten K., 1981, Peptides in metabolic autonomic nerves, Diabetologia 20: 337-342.

van Houten M. and Posner B.I., 1979, Insulin binds to blood vessels in vitro, Nature 282: 623-625.

van Houten M. and Posner B.I., 1981, Cellular basis of direct insulin action in the central nervous system, Diabetologia 20: 255-267.

van Houten M., Posner B.I., Kopriwa B.M. and Brawer J.R., 1980, Insulin binding sites localized to nerve terminals in rat median eminence and arcuate nucleus, Science 207: 1081-1083.

van Houten M., Posner B.I., Kopriwa B.M. and Brawer J.R., 1979, Insulin binding sites in the rat brain: in vivo localization to the circumventricular organs by quantitative radioautography, Endocrinology 105: 666-679.

van Ness J., Maxwell I.H. and Hahn W.E., 1979, Complex population of nonpolyadenylated messenger RNA in mouse brain, Cell 18: 1341-1349.

Villa-Komaroff L., 1980, The synthesis of eukaryotic proteins in prokaryotic cells, in Gene Structure and Expression, ed. D.H. Dean, L.F. Johnson, P.C. Kimball and P.S. Perlman, Ohio State University Press, Columbus, pp. 81-105.

Villa-Komaroff L., Efstratiadis A., Broome S., Lomedico P., Tizard R., Naber S.P., Chick W.L. and Gilbert W., 1978, A bacterial clone synthesizing proinsulin. Proc. Nat. Acad. Sci. USA 75: 3727-3731.

Wensink P.C., Finnegan D.J., Donelson J.E. and Hogness D.S., 1974, A system for mapping DNA sequences in the chromosomes of D. melanogaster. Cell 3: 315-325.

Weyhenmeyer J.A. and Fellows R.E., 1983, Presence of immunoreactive insulin in neurons cultured from fetal rat brain, Cell Molec. Neuro. 3: 81-86.

Woods S.C. and Porte D. Jr., 1977, Relationship between plasma and cerebrospinal fluid insulin levels of dogs, Amer. J. Physiol. 233: E331-E334.

Woods S.C., Lotter E.C., McKay L.D., and Porte Jr., D., 1979, Chronic intracerebroventricular infusion of insulin reduces food intake and body weight of baboons, Nature 282: 503-505.

Yallow R. and Eng J., 1983, Insulin in the central nervous system, Advances in Metabolic Disorders 10: 341-354.

Yang J.W., Raizada M.K. and Fellows R.E., 1981, Effects of insulin on cultured rat brain cells: stimulation of ornithine decarboxylase activity, J. Neurochem. 36: 1050-1057.

Young R.A. and Davis R.W., 1983, Efficient isolation of genes by using antibody probes, Proc. Natl. Acad. Sci. USA 80: 1194-1198.

Zabeau, M., and Roberts, R.J., 1979, The role of restriction endonucleases in molecular genetics, in Molecular Genetics, ed. J.H. Taylor, Academic Press, NY. pp. 1-63.

IMMUNOLOGY AND EMBRYOGENESIS:

THE CHROMOSOMAL EDITING HYPOTHESIS

William J. Dreyer and Janet M. Roman

Division of Biology
California Institute of Technology
Pasadena, CA   91125

INTRODUCTION

Time lapse movies of migrating T-cells remind us very much of
nerve growth cones. Both sense the environment with ameboid-like
processes as they seek their respective targets. The genetic,
molecular and cellular basis for the production of the specificity
on B-cells and T-cells is understood to an extraordinary degree at
this point in history. On the other hand, the specificity found on
cells in the developing nervous system is only understood at the
level of descriptive biology.

During embryogenesis of the immune system, a series of
systematic developmental decisions is made that generates an
enormous diversity of cells. Millions of different T-cells and
B-cells are formed during the process. Each individual cell line
is genetically committed to expression of only one receptor with
unique specificity. These receptors are the products of
chromosomal editing, which occurs at specific points as these cell
lineages develop. DNA is actually cut and spliced in a programmed
manner and daughter cells have an altered arrangement of specific
genes (Dreyer and Bennett, 1965; see Leder, 1982 for a general
review). This edited version of the chromosomal program is then
replicated in subsequent cell divisions within the lineage.

Principles of molecular evolution have led us to believe that
the sophisticated, programmed chromosomal editing mechanism found
in the developing immune system did not appear de novo in the first
vertebrates but evolved from a more general but similar mechanism
of programmed lineage determination. Chromosomal editing may help
explain embryogenesis in general, and the generation of highly
specific cell surface recognition processes in particular.

Chromosomal Editing of the Genes that Code for Antibodies

At appropriate stages of development both T- and B-cells display antibody-like receptors on their surfaces. The receptors, like circulating antibody molecules, are composed of disulfide bonded protein monomers. Each monomer is made up of two or more functional domains. The domains which form the antigen binding site are known as V or variable regions while the remaining portions of the receptors are referred to as C or constant regions. It is the constant region that interacts with the cell membrane and transduces a signal to the cell after the variable regions have interacted with the specific molecular structure against which they are targeted. A great deal is known about these molecules and the nature of their interactions. For example, not only is the complete amino acid sequence known for a large number of different antibody molecules but an enormous amount of DNA sequence is also known, both for the edited antibody genes from specific differentiated cells and for the original master copy genes from germ line cells (see Leder, 1982 for a general review).

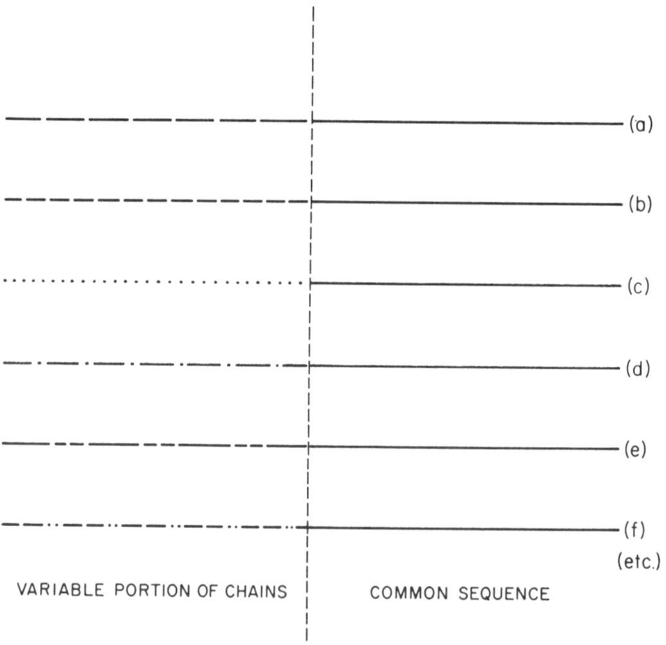

Fig. 1. Each line represents an antibody light chain molecule. The solid portion of each represents the constant (C) region and the broken portions represent different variable (V) regions. These molecules are coded for by genes that are the products of chromsomal editing events.

The general principles which are involved in the commitment
of cells to production of specific antibody light chains are shown
in Figure 1 (Dreyer and Bennett, 1965). Each line in the figure
can be taken to symbolize a different antibody light chain. The
amino acid sequence of the V-region is different in each of the
different antibody light chains, while the C-region is absolutely
identical. Analysis of data which was collected in this field
about two decades ago led to the realization of an extraordinary
phenomenon: the genes which code for antibody molecules are edited
during embryological development. They are cut and spliced so that
each daughter cell line has a different V-region gene spliced in
proximity to the same C-region gene. This newly edited DNA
sequence is then duplicated as the cell divides. This explains the
way in which each specific cell type in the lineage retains and
replicates the information needed to construct its particular cell
surface receptor or antibody molecule.

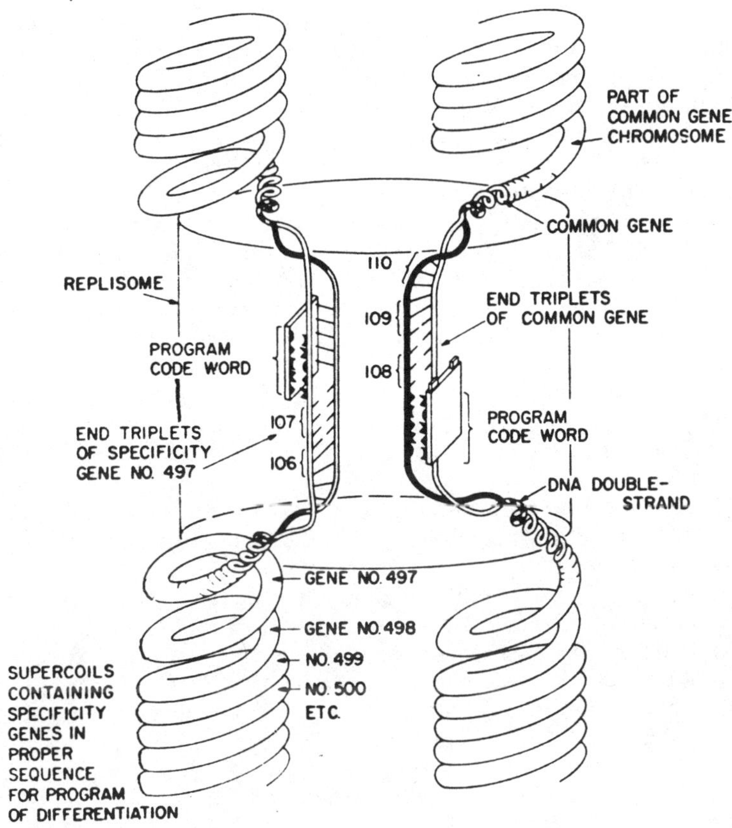

Fig. 2. Illustration of tandem arrays of V genes and control
sequences within a chromosome. Our hypothesis suggests that the
linear arrangement of genes helps program sequential gene editing
during development. See text.

Figure 2, taken from Dreyer and Gray, 1968, illustrates some points that have now been well established and suggests others that are still the subject of research and hence remain speculative. As postulated, the genes which code for V-regions (termed specificity genes in the figure) are illustrated lined up in tandem arrays on a specific chromosome, as shown on the left hand side of the drawing, e.g., genes 497, 498, 499, 500, etc. Control regions consisting of a relatively small number of base pairs were viewed as being adjacent to the coding sequences. It was assumed that these control regions were recognized in a very specific manner and that this recognition led to the cutting and splicing of specific antibody V-genes to the same constant (i.e., common) gene. This hypothesis has been shown to be basically correct. DNA sequencing has proven that sequences not adjacent in the embryo are cut and moved into a functional relationship in the adult.

The process of cutting and splicing of these genes was envisioned to be mediated by a protein-enzyme complex termed the <u>replisome</u>. The replisome is able to recognize specific splice signals (program code words) and edit the DNA. This editing process is now known to involve two steps. First a V-gene is spliced adjacent to one of a number of different J-genes. Then, in a second editing event, that combination is spliced near the C-region as a daughter cell is committed to production of a specific light chain protein. The process of editing of DNA has been monitored by examining the actual sequence of DNA in specific cloned cell lines. The enzyme complex (replisome) that is responsible for accomplishing the editing is still hypothetical.

An important hypothetical concept illustrated in Figure 2 is the possibility that the actual linear arrangement of V-genes within the multigene complex plays a role in programming the sequence in which they are expressed during development. Work with inbred chick embryos has provided experimental evidence which suggests that orderly programming of antibody production does occur (Huang and Dreyer, 1978). In chickens, the bursa is the site of differentiation of antibody producing cells. If generation of antibody diversity occurs in a precise order, the destruction of the bursa during development should result in a chicken with a limited repertoire of antibody diversity that is identical for all chicks bursectomized at the same time of development. By comparing complex patterns of antibody light chains separated electrophoretically on the basis of molecular weight and charge, Huang and Dreyer observed restricted patterns in bursectomized birds compared with normal birds. These patterns were increasingly complex the later in development bursectomy occurred, and more importantly, the patterns for birds bursectomized at a similar point in development were very similar, suggesting that the mechanisms generating antibody diversity function (at least initially) in a sequential order that is nearly identical in different embryos. Such experiments do not, however, address the question of whether or not this orderly production of specific antibodies is the direct result of a particular linear arrangement of genes on the chromosome. This question will undoubtedly be answered soon, as many laboratories are determining the DNA sequence of V-genes, and it should be possible to compare their spatial relationships on germ-line chromosomes with their expression during embryogenesis. Even with our current knowledge, it is very clear that programmed chromosomal editing in some form does contribute to cellular commitment to both heavy and light antibody chains and also to the T-cell receptors (Hedrick et al., 1984a, 1984b; Yanagi et al., 1984). Other more random mutational events are superimposed on the basic editing process and contribute to further diversity.

It should be emphasized that chromosomal editing of the type that occurs in the immune system is subtle. It was initially extremely hard to detect and has been localized to specific gene families on each of several different chromosomes. We propose that similarly subtle events occur in other lineages and that as a rule they will be equally difficult to detect. Nevertheless the means to study these events are becoming available (see Dreyer et al., this volume and Hunkapiller et al., 1984) and should allow these processes to be elucidated in the future.

Evolutionary Origins of Programmed Chromosomal Editing

We hypothesize that mechanisms similar to the sophisticated and programmed editing of genes that occurs during the generation of diversity in B- and T-cells may also be involved in the generation of cellular diversity in other organ systems. This hypothesis becomes very attractive when one examines existing data from the viewpoint of molecular evolution. Antibody molecules are found in the earliest vertebrates to appear during evolution, and yet there is no evidence of their occurrance in invertebrates. From what did these proteins and editing mechanisms evolve? The high degree of homology seen in the sequences of antibody molecules across species, combined with the degree and type of sequence divergence seen between the earliest and the latest vertebrates, strongly suggests that these molecules and mechanisms evolved from similar pre-vertebrate molecules rather than in an evolutionary "big bang" with the first vertebrates (Dreyer et al., 1967; Dreyer, 1984). A mechanism capable of programming the generation of large numbers of B- or T-cell receptors could also be capable of generating diversity in other receptors and would handily explain a wide variety of observed phenomena. Consider for example, cells of the neural crest that migrate to form melanocytes that "know" the address to which they are to go as they generate the genetically programmed pigment patterns on tropical fish. Other migrating neural crest cells form portions of cartilage or bone that are assembled in precisely sculptured forms, again under genetic command. Still other cells from the neural crest form neurons which send processes to genetically predetermined targets so that the system can be wired up accurately. We feel that molecules very much like the B-cell or T-cell receptors could help explain numerous phenomena concerning the cell migrations and cell associations that occur throughout embryogenesis.

We propose, therefore, that chromosomal editing mechanisms similar to those that splice the genes coding for the B- and T-cell receptors could generate not only one but a series of different receptors that constitute a surface display on a single cell (Fig. 3). For example, one receptor might sense anterior regions, a second receptor ventral positions, and a third a specific compartment in the anterior ventral portion of the embryo. A relatively small number of additional receptors on each cell could

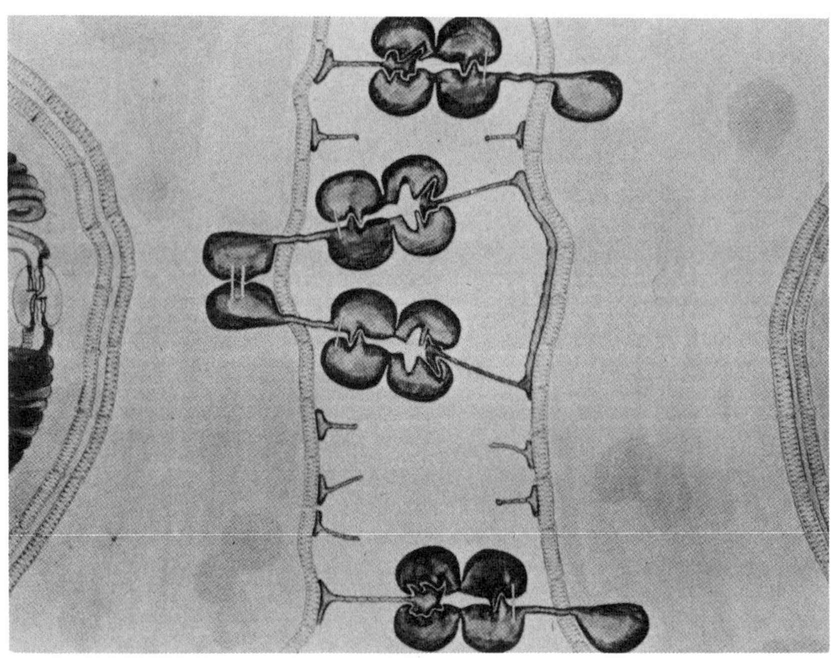

Fig. 3. A diagramatic illustration of the specific interactions of hypothetical cell-surface molecules.

help determine the relationships among cells with a very high degree of detail and sophistication. According to this hypothesis, cells display a number of discrete address-sensing molecules on their surfaces yet they might appear to move as if in a gradient as they seek out a correct address (see Hood et al., 1977; Dreyer, 1984).

The chromosomal editing hypothesis, in addition to providing a mechanism for generating spatial organization by means of specific cell receptors, also provides a mechanism for generating temporal organization so that the events in a developing organism occur in a sequential, orderly way. This occurs as genes are sequentially edited during development, and in this way cells are assigned to particular lineages. While the chromosomal editing hypothesis predicts that stem cells become increasingly restricted in potential as differentiation proceeds, this process does not compromise the potential of the animal as a whole. This is because of the presence of residual, redundant stem cells (see below) which have not yet undergone genetic editing.

The nuclear transplantation studies of Gurdon (e.g.: Gurdon and Laskey, 1970), in which nuclei from differentiated cells give rise to at least some normal developmental processes when transplanted into enucleated eggs, have often been put forward as evidence against chromosomal changes during development. However, we know beyond any doubt that genes are edited in mature lymphocytes, and yet nuclei from such cells can undergo development up to the early tadpole stage (Wabl et al., 1975). It is noteworthy, however, that no proven differentiated nucleus has ever given rise to an adult (McKinnell, 1978). Indeed, differentiated nuclei do not develop at all if transplanted directly, but require multiple serial passages during which recombination could occur with nonedited alleles. We wish to emphasize two points: (1) The chromosomal editing in lymphocytes (and thus the editing we predict for other tissues) is very subtle, involving a minute fraction of the entire genome. (2) Unless one assumes that the failure of differentiated nuclei to develop beyond the tadpole stage is due to chromosomal editing, one must conclude that the technique of nuclear transplantation is simply not a satisfactory experimental technique to detect or analyze the kinds of chromosomal changes that occur in lymphocytes.

The chromosomal editing hypothesis, while relying on somatic-genetic events, nevertheless requires the interaction of a variety of epigenetic factors and stimuli to mediate embryogenesis. In the immune system, for example, the generation of diversity in T- and B-cells is programmed by precise gene cutting and splicing, but epigenetic factors of many kinds influence these events. Cell-cell and cell-tissue interactions are crucial. A myriad of hormones and growth factors provide essential signals at various developmental

levels. Certain cell-cell interactions suppress development of
particular lineages, others stimulate. Antigens, too, can either
suppress (immune tolerance) or stimulate development. Antigens
also induce additional highly specific gene splicing events (e.g.,
heavy chain class switching) and can stimulate extensive cell
differentiation without cell division. Other epigenetic factors
include, for example, high or low levels of Ca^{++} or cyclic AMP. We
will discuss some of the implications of the chromosomal editing
hypothesis below.

**Chromosomal Editing Must Occur in Both Regulative and Mosaic
Development**

Chromosomal editing forms a framework in which to understand
two apparently contradictory modes of development, mosaic versus
regulative.

The nematode is an example of an organism that undergoes
rigid mosaic development. Each cell generated from the dividing
fertilized egg is committed to a specific cell lineage in the
adult. Although laser ablation experiments indicate that in some
cases cells of one lineage may develop into those of another, such
examples are unusual and involve cells which are already very
similar (Horvitz, 1981). Each cell division in the nematode
results in new cells with increasingly restricted developmental
options. Destruction of a cell at one specific stage of
development generally leads to the permanent loss of the entire
portion of the organism which would have been generated by that
specific cell. If the cell is one of a small number of cells in an
early embryo, a major deficit is seen. If it is a cell with only
one or two divisions remaining before its developmental pathway is
complete, only a very small but specific group of cells is lost
from the final organism. Somatic cell regeneration is rare in
nematodes; cells appear to be essentially hard wired in a
genetically predetermined pattern. In fact, many nematodes
actually undergo microscopically visible gross chromosomal editing
in somatic cells with only germ cells exempted (see for example,
Boveri, 1904).

Other organisms such as fish and amphibians undergo
regulative development. In this form of development, in contrast
to the nematode, cells can be replaced after being destroyed. Even
entire limbs can be regenerated, complete with properly functioning
sensory and motor neurons. For example, a lizard whose tail has
been pulled off by a predator is capable of regenerating a
completely new one.

How can we explain what appear to be such profound
differences between mosaic and regulative development? We believe
the explanation lies in the fact that organ systems which are

capable of regulation and regeneration have stem cells which are
both <u>redundant</u> and <u>residual</u>.

Redundant and Residual Stem Cells

Organisms which are capable of undergoing regulative
development produce multiple redundant stem cells each capable of
giving rise to the identical cell lineage upon stimulation. Those
residual stem cells which do not participate in the initial
differentiative events nevertheless remain viable and capable of
regenerating that specific cell lineage upon its loss. We will
illustrate the behavior of redundant and residual stem cells by
describing the immune system.

The cell lineage that ultimately forms both red and white
blood cells (including the B-cells and T-cells of the immune
system) includes stem cells capable of both generating and
regenerating each of these cell types. Daughter stem cells are
produced by the more pluripotent stem cell, and each of these new
lineages is then irreversibly restricted to the generation of a
specific subset of blood cells. For example, there are T-cell
precursors that migrate to the thymus where they continuously
generate an enormous diversity of T-cells but can never be induced
to revert to form stem cells of the red blood cell lineage. Stem
cells in the thymus represent a redundant system in that there are
many identical stem cells in the organism and each is capable of
generating a large family of daughter T-cells. Most of the T cells
generated each day ($>10^7$ cells) are not needed or used and die in
the thymus. Only a small number leave the thymus to carry out
their highly specific functions. There are also residual stem
cells that are retained and replicated in the thymus. These
residual stem cells allow the thymus to continuously regenerate
daughter T-cells. Redundant, residual stem cells also function in
the formation of red blood cells. These stem cells are found in
the bone marrow and continuously regenerate the various cells in
the red cell lineage.

The nematode has no redundant stem cells in the lineages
which form its somatic tissues. Furthermore, it generally leaves
no residual stem cells behind; when daughter cells are produced, no
potential mother cells of that type remain. The germ cell line of
the nematode represents an exception since both redundant and
residual germ cells are formed, which explains the ability of
nematodes to regulate the production of germ cells and to
regenerate them indefinitely. Conversely, it is the absence of
both residual and redundant stem cells specific for somatic tissues
that explains the rigid mosaic development, and the absence of
regulation and regeneration, seen in the nematode.

SUMMARY

We have elaborated the chromosomal editing hypothesis of development. This hypothesis, based on evolutionary arguments, states that the immune system must have evolved from pre-existant cell receptor systems on other tissues. Therefore we feel that what we know about development in the immune system can serve as a provisional model for studying development in other organ systems. The model predicts that specifically programmed somatic-genetic events occur as lineages develop. These DNA cutting and splicing events, similar to those which generate antibody molecules, generate a wide diversity of specific cell receptors which allow the cell to properly orient itself in the developing organism. In addition, the chromosomal editing events occur sequentially and generate a temporal organization so that developmental events occur in an orderly fashion as lineages develop. At any point in development, epigenetic factors play a crucial role in triggering cells to undergo specific differentiative events. We believe that this hypothesis explains the apparent contradiction between rigid mosaic development and regulative development by proposing that the same general types of somatic-genetic and epigenetic events occur in both. In mosaic development, however, the organism does not have the residual and redundant stem cells which allow for the cell replacement, repair and regeneration seen in regulative development.

We feel that by drawing analogies as we have from the development of the immune system one may gain insights into the genetic and epigenetic factors which govern the development of other tissues and formulate specific experiments to test the resulting hypothesis.

REFERENCES

Boveri, T. (1904). Ergebnisse uber die Konstitution der chromatischen Subotany des Zellkerns. Verlag von Gustav Fischer in Jena.

Dreyer, W. J. and Bennett, J. C. (1965). The molecular basis of antibody formation: A paradox. Proc. Natl. Acad. Sci. USA **54**: 864-869.

Dreyer, W. J., Gray, W. and Hood, L. (1967). The genetic molecular, and cellular basis of antibody formation: Some facts and a unifying hypothesis. Cold Spring Harbor Symp. Quant. Biol. **32**: 353-367.

Dreyer, W. J. and Gray, W. R. (1968). On the role of nucleic acids as genes conferring precise chemospecificity to differentiated cell lines. : "Nucleic Acids in Immunology," O. J. Plescia and W. Braun, eds. Springer-Verlag, New York, pp. 614-643.

Dreyer, W. J. (1984). Molecular evolution antibody formation and embryogenesis. In: "The Impact of Protein Chemistry on The Biomedical Sciences," A. N. Schechter, A. Dean, R. F. Goldberger, eds. Academic Press, New York, pp. 137-157.

Gurdon, J. B., and Laskey, R. A. (1970). The transplantation of nuclei from single cultured cells into enucleate frogs eggs. J. Embryol. Exp. Morphol. **24:** 227-248.

Hedrick, S. M., Cohen, D. I., Nielsen, E. A., and Davis, M. M. (1984a). Isolation of cDNA clones encoding T cell-specific membrane-associated proteins. Nature **308:** 149-153.

Hedrick, S. M., Nielsen, E. A., Kavaler, J., Cohen, D. I., and Davis, M. M. (1984b). Sequence relationships between putative T cell receptor polypeptides and immunoglobulins. Nature **308:** 153-158.

Hood, L., Huang, H. V. and Dreyer, W. J. (1977). The area code hypothesis: The immune system provides clues to understanding the genetic and molecular basis of cell recognition during development. J. Supramol. Struct. **7:** 531-559.

Horvitz, R. H. (1981) Neuronal Cell Lineages in the Nematode Caenorhabditis elegans In: "Readings in Developmental Neurobiology" Patterson, P. H. and Purves, D., Eds.   Cold Spring Harbor Laboratory. p. 140-155.

Huang, H., and Dreyer, W. J. (1978). Bursectomy in ovo blocks the generation of immunoglobulin diversity.   J. Immunol. **121:** 1738-1747.

Hunkapiller, M., Kent, S., Caruthers, M., Dreyer, W., Firca, J., Giffin, C., Horvath, S., Hunkapiller, T., Tempst, P., and Hood, L. (1984). A microchemical facility for the analysis and synthesis of genes and proteins. Nature (in press).

Leder, P. (1982). The genetics of antibody diversity. Scientific American **246:** 102-115.

McKinnell, R. G. (1978).   "Cloning: Nuclear Transplantation in Amphibia" University of Minnesota Press. p. 134-147.

Wabl, M. R., Brun, R. B., and DuPasquier, L. (1975) Lymphocytes of the toad Xenopus laevis have the gene set for promoting tadpole development. Science **190:** 1310-1312.

Yanagi, Y., Yoshikai, Y., Leggett, K., Clark, S. P., Alksander, I., and Mak, T. W. (1984). A human T-cell-specific cDNA clone encodes a protein having extensive homology to immunoglobulin chains. Nature **308:** 145-149.

GENETIC CONTROL OF THE CONNECTIVITY AND EXCITABILITY OF CEREBELLAR

PURKINJE CELLS IN RODENTS

Francis Crepel, Jean-Luc Dupont and Robert Gardette

INSERM U97

2ter, rue d'Alésia, 75014 Paris, France

The development of synaptic contacts between nerve cells and of the excitability of neuronal membranes are two important steps in the building of the nervous system. Hughes (1968a,b) was the first to state that during the normal course of development, connections are initially more diffuse than in the adult, i.e. the adult-type neuronal circuits derive not only from an increase in the total number of synaptic contacts, but also from the elimination of irrelevant connections. Indeed, this idea was not really new since at the turn of the century, Cajal (1911) had already described the presence of transient axonal projections in the central nervous system of mammals at early developmental stages.

However, the first demonstration of a regression of functional synapses during development only occured in 1970, when Redfern showed that, at the neuromuscular junction, the adult one-to-one relationship between motor axons and muscle fibres is preceeded in the neonate by a multiple innervation of motor end plates. This discovery subsequently gave rise to number of experiments (which will not be considered here) to analyse the possible mechanisms underlying this regressive phenomenon (see review in Brown et al., 1976 ; Ding et al., 1983).

A further advance in this field came when Changeux et al. (1973, 1976) and Gouzé et al. (1983) proposed that the development of the nervous system necessarily involves a selective stabilization of synaptic contacts, governed by the activity of the neuronal network. Thus, during development, a certain degree of specificity of nerve cell connections would be achieved by a matching of pre- and postsynaptic partners, dependent either on the presence of

specific and complementary chemical labels carried by neuronal
membranes (Sperry, 1963 ; Meyer and Sperry, 1976), or on the
presence of cytochemical gradients of such labels (Von der Malsburg
and Willshaw, 1977). Later on, a further shaping of the connectivity
would result from the selective stabilization of these performed
synaptic contacts by the traffic of impulses within the developing
nervous system.

Since the demonstration that a regression of already functional
synapses does occur in the normally developing central nervous system
is an obvious prerequisite to accept such theoretical views, the
first part of this review will deal with experiments which
demonstrated for the first time that this is indeed the case during
the establishment of the adult climbing fibre - Purkinje cell
relationships in the cerebellum of rodents. The possible causes
of these involutive processes will be also considered, as well as
the role of this synaptic remodeling in the postnatal shaping of
the olivocerebellar circuits. Other examples of transient
connections during development of the nervous system have been
described since then, for instance in the submandibular ganglion
of the rat (ref. in Purves, 1983) and among callosal projections
(Innocenti and Caminiti, 1980 ; Koppel and Innocenti, 1983), thus
extending experimental evidence in favor of the hypotheses quoted
above. However, so far, a role of the activity of sets of neurones
in these synaptic reorganizations remains to be determined.

The development of the excitability of neuronal membranes
is another important issue in developmental neurobiology. At the
moment, it is already well established that in invertebrates as
well as in vertebrates, ionic conductances appear either in a
sequential order during development (see for instance Llinas and
Sugimori, 1979 ; Spitzer, 1982 ; Salkoff and Wyman, 1983), or
simultaneously (Schwartzkroin, 1981 ; Peacock and Walker, 1983)
depending on cell types. This raises, among others, the question
of the genetic control of the development of ionic channels bourn
by excitable cells. In invertebrates as in unicellular animals,
a genetic dissection of ionic conductances has been already
achieved for sodium (Na), potassium (K) and Calcium (Ca) conductances
(Kung et al., 1975 ; Jan et al., 1977 ; Wu and Ganetzky, 1980 ;
Salkoff and Wyman, 1983). In the second part of this paper, a first
example of a selective alteration of a voltage-dependent ionic
conductance by a point mutation in vertebrate neurones will be
briefly described, and some possible causes of this defect will
be considered.

100

MULTIPLE INNERVATION OF PURKINJE CELLS IN THE DEVELOPING CEREBELLUM

Evidence for a Multiple Innervation

In the adult, responses elicited in Purkinje cells by the
activation of climbing fibres are very stereotyped (Fig. 1E,1F):
they consist of an initial fast spike followed by a plateau of
depolarization (10-20 msec in duration) on which partial spikes

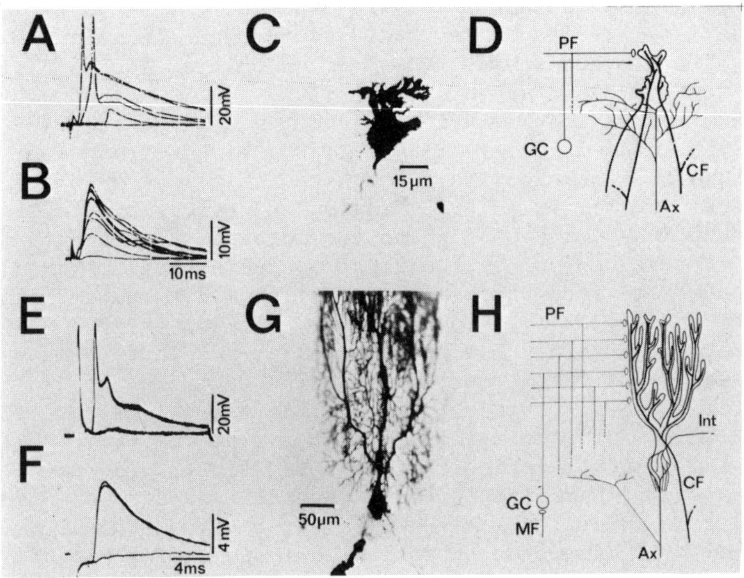

Fig. 1. Multiple innervation of Purkinje cells by climbing fibres.
A : climbing fibre response (intracellular recording)
elicited in a 7 day old rat by stimulation of the cerebellar
white matter at various intensities. B : immature climbing
fibre EPSP in a 6 day old rat. Note the graded character
of the responses in A and B. C : multiply innervated
Purkinje cell in a 8 day old rat, stained by intracellular
injection of HRP. D : afferent connections to Purkinje
cells in the immature rat : Ax : axone, CF : climbing
fibres, GC : granule cells, PF : parallel fibres. E,F :
all-or-none climbing fibre response (E) and EPSP (F) in
an adult rat. G : adult Purkinje cell stained with HRP.
H : afferent connections to Purkinje cells in the adult
rat : Int : interneurone, MF : mossy fibres, Ax, CF, GC,
PF : as in D.

are superimposed. Furthermore, they are all-or-none in character,
due to one-to-one relationships between climbing fibres and Purkinje
cells (Eccles et al., 1967). When recorded extracellularly, these
responses are also easily identified as all-or-none bursts of
spikes, the so-called "complex responses" (Thach, 1968).

The electrophysiological demonstration of a multiple
innervation of Purkinje cells by climbing fibres in the immature
cerebellum of the rat was initially done on postnatal day 7 to 9
(Crepel et al., 1976). It was based upon the observation that
climbing fibre responses and EPSPs recorded from Purkinje cells
at this developmental stage differ in one important respect from
those observed in the adult in that they vary in a stepwise manner
with the stimulus intensity (Fig. 1A,1B), rather than being
all-or-none in character. This suggested that in the immature rat,
every Purkinje cell is multiply innervated by climbing fibres, each
step in the EPSP corresponding to the activation of an additional
fibre. Since at the end of the first postnatal week, Purkinje cells
can also be activated through mossy fibres and granule cells
(Woodward et al., 1971 ; Crepel, 1974 ; Shimono et al., 1976),
complementary experiments were required to rule out the possibility
that these graded responses represent the summation of excitatory
effects exerted on Purkinje cells by mossy as well as by climbing
fibres, which would avoid the necessity of postulating a multiple
innervation of Purkinje cells by climbing fibres. Accordingly,
the responses of Purkinje cells to cerebellar afferents were
analysed in rats of the same age, in which either the climbing
fibres had been disrupted by extensive lesions of their unique
source, the inferior olive, or the mossy fibre - granule cell
pathway had been interrupted by X-irradiation of the pups, a
procedure which destroys all the cerebellar granule cells. In the
former group, the responses interpreted in normal animals as due to
the activation of Purkinje cells by climbing fibres were no longer
present, whereas they persisted in degranulated cerebella. Altogether,
these results therefore demonstrated that at this developmental
stage, Purkinje cells are multiply innervated by climbing fibres
(Fig. 1D) (Crepel et al., 1976).

Similar experiments were performed at different developmental
stages to study the normal course of the multiple innervation (Crepel
et al., 1981 ; Mariani and Changeux, 1981a,b). In particular, Crepel
et al. (1981) performed control experiments on X-irradiated animals
at all developmental stages studied, to ascertain that the responses
recorded in normal pups were entirely due to climbing fibre input
to Purkinje cells. It was shown that, typically, Purkinje cells
are contacted by climbing fibres as early as day 2 or 3, and rapidly
become innervated by, on average, 3 or 4 fibres on day 5. Later on,
a rapid regressive process occurs, leading by day 13 to the adult
pattern of innervation (Fig. 2).

102

The comparison of the normal course of multiple innervation
with that observed in X-irradiated rats during development (Crepel
et al., 1981) not only validated the demonstration of supernumerary
climbing fibre projections onto Purkinje cells in immature animals,
but also brought insights into the mechanisms responsible for their
regression. It is apparent from Fig. 2B that the early phase of
the regressive process is identical in both groups of animals. In
contrast, the late phase of the regression of the multiple
innervation (after day 8) does not occur in X-irradiated animals,

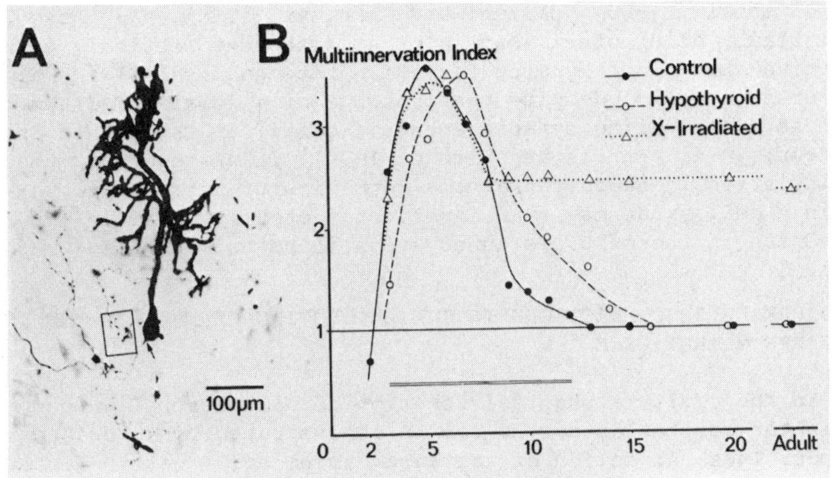

Fig. 2. A : Purkinje cell from an adult X-irradiated rat.
Intracellular staining with HRP. B : evolution of the mean
number of climbing fibres per Purkinje cell versus age in
control and experimental rats.

leading to the persistance of the multiple innervation in the
adult. Indeed, Woodward et al. (1974) previously demonstrated the
presence of a multiple innervation of Purkinje cells by climbing
fibres in adult X-irradiated rats. From these results, we proposed
that the early phase of the regression of the multiple innervation
of Purkinje cells by climbing fibres is due to a homosynaptic
competition among climbing fibres to innervate Purkinje cells, as
occurs at the neuromuscular junction or in the submandibular
ganglion. In contrast, the late phase of the regression seems to
be dependent on the formation of synapses between parallel fibres
(the axons of granule cells) and Purkinje cells, possibly because
the formation of these synapses strengthen competition among
climbing fibres by decreasing the supply of trophic factors emitted
by Purkinje cells.

Other evidence supporting this role of granule cells came from electrophysiological experiments on climbing fibre - Purkinje cell relationships in cerebellar mutant mice. In particular, it was shown that, in the agranular cerebellum of weaver mutant mice, all Purkinje cells are multiply innervated by climbing fibres (Crepel and Mariani, 1976). In reeler mutant mice, this is also true for abnormally located Purkinje cells, i.e. cells lying near cerebellar nuclei and therefore devoid of parallel fibre inputs (Sotelo, 1980) whereas normally located Purkinje cells are typically monoinnervated by climbing fibres (Mariani et al., 1977). Finally, a multiple innervation of Purkinje cells by climbing fibres was also demonstrated in adult staggerer mutant mice (Fig. 3) (Crepel et al., 1980; Mariani and Changeux, 1980), a mutation which leads, among other abnormalities (see next section), to a selective absence of synapse formation between parallel fibres and Purkinje cells, despite the presence of a substantial number of these axons during synaptogenesis (Sidman, 1972; Sotelo and Changeux, 1974; Landis and Sidman, 1978). Taken together, these results strongly suggest that the formation of parallel fibre - Purkinje cell synapses is an important factor involved in the regression of the multiple innervation in normal animals (Crepel, 1982).

OLIVOCEREBELLAR CONNECTIONS IN THE DEVELOPING CEREBELLUM AND IN STAGGERER MUTANT MICE

In the adult, a sagittal zonation of olivocerebellar connections within the cerebellar cortex can be demonstrated by autoradiographic methods. Thus, injection of tritiated amino acids within discrete zones of the inferior olive leads to an anterograde transport of

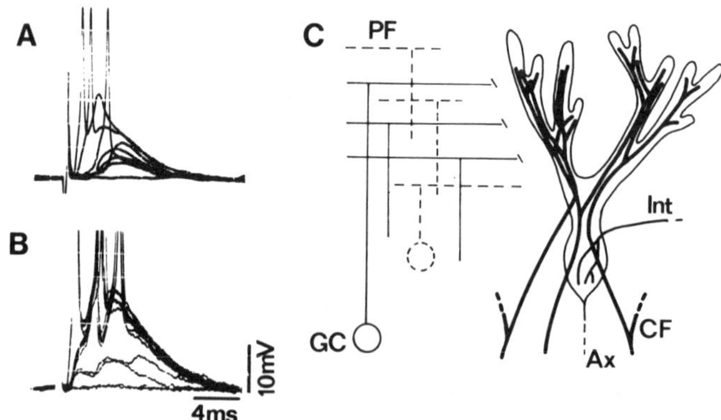

Fig. 3 : A,B: intracellular recordings from 2 Purkinje cells in a staggerer mutant mouse. The climbing fibre responses were graded by steps with the intensity of the stimulus. C: cerebellar circuits in staggerer mouse.

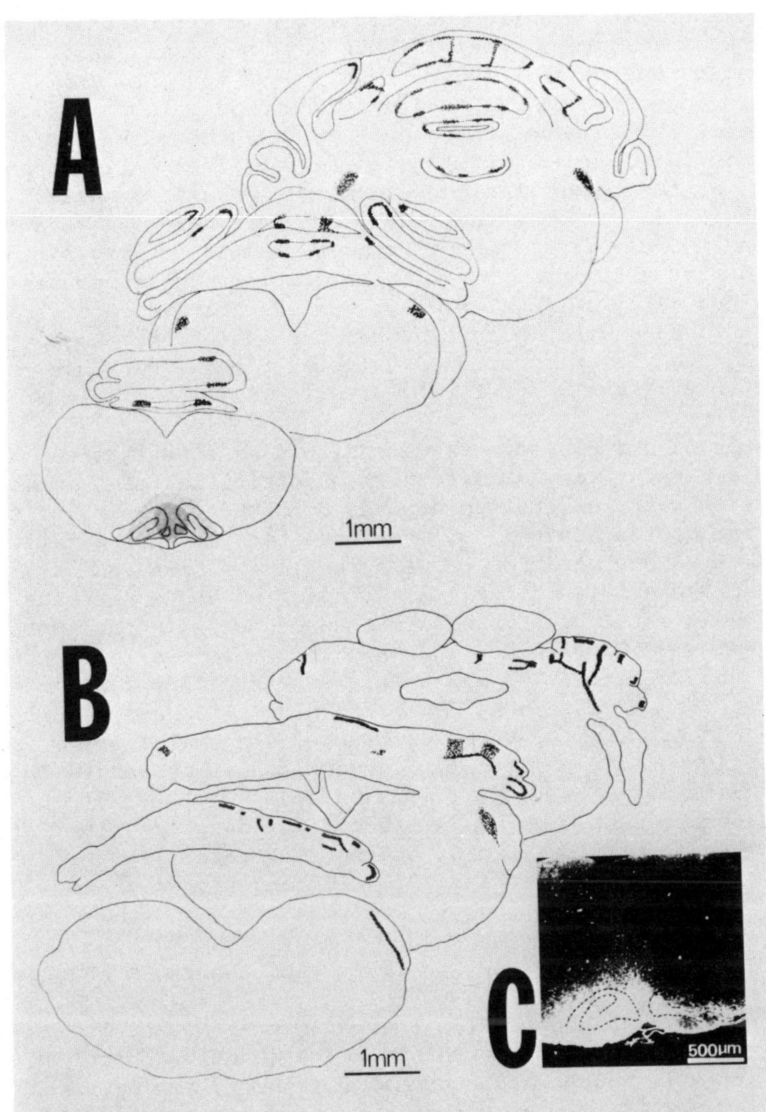

Fig. 4. A,B : camera lucida drawings of cerebellar sections from a 7 day old rat (A), and from an adult staggerer mutant mouse (B) showing the distribution of labeled climbing fibres within the cerebellar cortex. C : darkfield microphotograph of the injection site of 3H-Leucine in the inferior olivary complex of the adult staggerer mouse. The injection site is also visible on the most caudal section of the medulla oblongata of the 7 day old rat (A).

radioactive material up to the terminal of climbing fibres which are
distributed in sagittal bands within the contralateral cerebellar
cortex (Courville, 1975 ; Groenewegen and Voogd, 1977 ; Campbell and
Armstrong, 1983). Could such a zonation result from an initially more
diffuse distribution of climbing fibres through the regression of the
multiple innervation during development ? To answer this question,
similar autoradiographic experiments were performed on 5 to 13 day
old rats (Dupont et al., 1981). Even in the youngest animals studied,
i.e. when injection of 3H-Leucine and autoradiographic processing of
the cerebella were done at the peak or near the peak of the multiple
innervation, a clear sagittal zonation of silver grain deposits was
apparent in the cerebellar cortex (Fig. 4A), with a maximum staining
around Purkinje cells, where climbing fibres make synapses with these
neurones at early developmental stages. From these experiments, it
appeared that a sagittal zonation of olivocerebellar connections is
established early in development. However, further studies are
required to determine if any further refinement of this distribution
occurs during the epigenesis of climbing fibre - Purkinje cell
relationships (Sotelo et al., 1984).

Autoradiographic methods were also used in staggerer mutant
mice to determine to what extent the distribution of climbing
fibres within the cerebellum depends upon the integrity of the
cerebellar cortex. Indeed, as mentioned before, this mutation
prevents synapse formation between parallel fibres and Purkinje
cells and leads to a multiple innervation of these neurones by
climbing fibres. Furthermore, a marked decrease in the number
of large to medium size neurones within the cerebellar cortex
(Herrup and Mullen, 1979) and a late degeneration of granule
cells (Sotelo and Changeux, 1974 ; Sotelo, 1980) are also apparent
in this mutant. Despite these pronounced abnormalities, a sagittal
organization of olivocerebellar connections is preserved in this
mutant (Fig. 4B,4C) (Dupont et al., 1983). Together with previous
autoradiographic studies on immature rats, these results suggest
that the sagittal organization of the olivocerebellar connections
is, at least in part, intrinsic to climbing fibres.

BIOELECTRICAL PROPERTIES OF PURKINJE CELLS IN STAGGERER MICE

Cerebellar Purkinje cells from staggerer mutant mice seem to
be particularly appropriate to study the genetic control of ionic
conductances in vertebrate neurones for the following reasons.
First, in normal animals, Purkinje cells exhibit several types of
ionic conductances including fast Na conductances at a somatic
level and Ca channels giving rise to well developed Ca spikes in
their dendrites (Llinas and Sugimori, 1980a,b). Secondly, anatomical
evidence suggests that Purkinje cells are directly affected by the
staggerer mutation and that most of the defects concern their
dendrites (Sotelo and Changeux, 1974). Thirdly, cerebellar slices

can be easily maintained in vitro and thus allow stable intracellular recordings from Purkinje cells as well as the bath application of drugs. We therefore decided to investigate the bioelectrical properties of these neurones in cerebellar slices maintained in vitro, as compared to those in normal mice, with special emphasis on Na and Ca conductances (Crepel et al., 1984).

The major finding of this study was that, in staggerer mice, Purkinje cells lack Ca spikes or any response which could be ascribed to the activation of Ca channels, whereas such responses are routinally observed in normal Purkinje cells (Fig. 5A1,5B1). In particular, this was true under conditions which normally enhance Ca currents, i.e. when K conductances were depressed by adding Tetraethylammonium (TEA) in the bathing medium, or when Ca was replaced by Barium (Ba) which, in addition to its depressant effect on K conductances, also passes more easily than Ca through

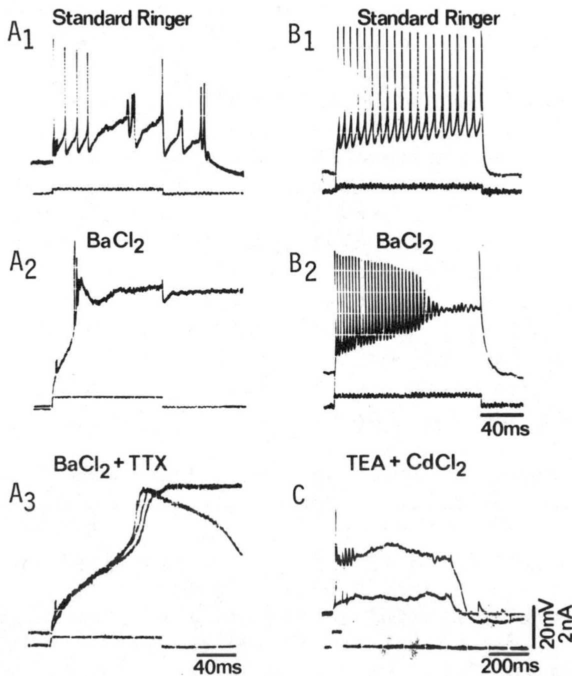

Fig. 5. Responses of Purkinje cells to direct electrical stimulation
in control (A) and staggerer mice (B,C) : upper traces :
membrane potentials ; lower traces : stimulating currents.
Note that the prolonged action potentials elicited in
normal Purkinje cells under Barium (Ba) are Tetrodotoxin (TTX)
resistant. Note also, in staggerer mice, the presence of
prolonged action potentials which persist in the presence of
of the Ca channel blocker Cadmium (Cd). These prolonged
Na-dependent spikes were abolished by TTX (not illustrated).

Ca channels, thus causing in normal Purkinje cells prolonged action
potentials which were not observed in staggerer (Fig. 5A2,5A3,5B2).
In contrast, fast Na-dependent action potentials and slowly inacti-
vating Na conductances were apparently unaffected by the mutation
(Fig. 5B1,5C).

Several reasons might account for the absence of Ca spikes in
staggerer Purkinje cells as, for instance, an abnormally low
density of Ca channels, or abnormal activation properties of these
channels, due for instance to a high internal concentration of free
Ca which would reduce Ca conductance (Hagiwara, 1983).

Whatever the cause of the defect is, several other questions
arose from these results. The first one was to determine to what
extent the absence of Ca spikes in staggerer mice is the consequence of the abnormal wiring of Purkinje cells in these animals. This
hypothesis was tested in reeler mutant mice since, as mentioned
before, in the cerebellum of this mutant Purkinje cells which are
normally located in the cortex are normally afferented by climbing
and parallel fibres, whereas cells which lie near the nuclei are
devoid of synaptic contacts with parallel fibres and are multiply
innervated by climbing fibres, i.e. a situation similar to that
observed in staggerer mice. Indeed, normally located Purkinje cells
as well as those lying in the central mass exhibited well developed
Ca conductances. Furthermore, in the latter, the distribution of
the Na and Ca channels was also apparently preserved since
Na-dependent spikes were much more prominent at a somatic level
than in the dendrites, whereas the inverse was true for Ca spikes
(Dupont et al., 1983b).These results suggest in turn that, in
staggerer mice, the absence of Ca spikes is not due to the abnormal
wiring of Purkinje cells but, rather, is a direct consequence of
the mutation on these neurones.

Another question was to determine the extent to which the
staggerer mutation affects Ca conductances in other classes of
neurones in the central nervous system. Since CA1 pyramidal cells
in the hippocampus normally exhibit well developed Ca conductances
(Schwartzkroin and Slansky, 1977), the bioelectrical properties of
these neurones were studied in staggerer mice as compared to normal
mice. In the mutant, bioelectrical responses of CA1 neurones to
stimulation of the afferents and to direct stimulation of the cells
were identical to those seen in controls. In particular, well
developed responses mediated through Ca channels were apparent in
the mutant (Fournier and Crepel, unpublished results). Therefore,
this again suggests that the impact of the staggerer mutation on
excitable properties of nerve cells is rather selective, i.e.
restricted to cerebellar Purkinje cells.

Finally, it must be emphasized that even if the defect in Ca
conductances observed in staggerer Purkinje cells is a remote
consequence of the primary effect of the mutation, it certainly
may have important consequences on the development and the function
of these neurones, given the key role attributed to Ca in metabolic
processes.

ACKNOWLEDGEMENTS

The authors wish to thank M.W. Debono for skilful technical
assistance and the Ministère de l'Industrie et de la Recherche
for supporting this work (grant n°83C0910).

REFERENCES

Brown, M.C., Jansen, J.K.S., and Van Essen, D. (1976). Polyneu-
     ronal innervation of skeletal muscle in new-born rats
     and its elimination during maturation. J. Physiol.
     (Lond.) 261:387-422.

Campbell, N.C., and Armstrong, D.M. (1983). Topographical loca-
     lization in the olivocerebellar projection in the rat :
     an autoradiographic study. Brain Res. 275:235-249.

Changeux, J.P., Courrège, P., and Danchin, A. (1973). A theory
     of epigenesis of neuronal networks by selective stabi-
     lization of synapses. Proc. Natl. Acad. Sci. (USA)
     70:2974-2978.

Changeux, J.P., and Danchin, A. (1976). Selective stabilization
     of developing synapses as a mechanism for the specifi-
     cation of neuronal networks. Nature 264:705-712.

Courville, J. (1975). Distribution of olivocerebellar fibers
     demonstrated by radioautographic tracing methods.
     Brain Res. 95:253-263.

Crepel, F. (1974). Excitatory and inhibitory processes acting
     upon cerebellar Purkinje cells during maturation in
     the rat ; influence of hypothyroidism. Exp. Brain Res.
     20:403-420.

Crepel, F. (1982). Regression of functional synapses in the
     immature mammalian cerebellum. Trends in Neurosci.
     5:266-269.

Crepel, F., Delhaye-Bouchaud, N., and Dupont, J.L. (1981). Fate
     of the multiple innervation of cerebellar Purkinje cells
     by climbing fibers in immature control, X-irradiated
     and hypothyroid rats. Dev. Brain Res. 1:59-71.

Crepel, F., Delhaye-Bouchaud, N., Guastavino, J.M., and Sampaio,
     I. (1980). Multiple innervation of cerebellar Purkinje
     cells by climbing fibres in staggerer mutant mouse.
     Nature 183:483-484.

Crepel, F., Dupont, J.L., and Gardette, R. (1984). Selective
        absence of calcium spikes in Purkinje cells of staggerer
        mutant mice in cerebellar slices maintained in vitro.
        J. Physiol. (Lond.) 346:in press.
Crepel, F., and Mariani, J. (1976). Multiple innervation of
        Purkinje cells by climbing fibres in the cerebellum
        of the weaver mutant mouse. J. Neurobiol. 7:579-582.
Crepel, F., Mariani, J., and Delhaye-Bouchaud, N. (1976).
        Evidence for a multiple innervation of Purkinje cells
        by climbing fibres in the immature rat cerebellum.
        J. Neurobiol. 7:567-578.
Ding, R., Jansen, J.K.S., Laing, L.G., and Tonnesen, H. (1983).
        The innervation of skeletal muscles in chickens
        curarized during early development. J. Neurocytol.
        12:887-919.
Dupont, J.L., Delhaye-Bouchaud, N., and Crepel, F. (1981).
        Autoradiographic study of the distribution of olivo-
        cerebellar connections during the involution of the
        multiple innervation of Purkinje cells by climbing fibers
        in the developing rat. Neurosci. Letters 26:215-220.
Dupont, J.L., Gardette, R., and Crepel, F. (1983a). Olivocere-
        bellar projections in control and staggerer mutant mice.
        Brain Res. 270:330-334.
Dupont, J.L., Gardette, R., and Crepel, F. (1983b). Bioelec-
        trical properties of cerebellar Purkinje cells in reeler
        mutant mice. Brain Res 274:350-353.
Eccles, J.C., Ito, M., and Szentagothai, I. (1967). "The
        cerebellum as a neuronal machine." Springer, Berlin.
Gouzé, J.L., Lasry, J.M., and Changeux, J.P. (1983). Selective
        stabilization of muscle innervation during development :
        a mathematical model. Biol. Cybern. 46:207-215.
Groenewegen, H.J., and Voogd, J. (1977). The parasagittal zona-
        tion within the olivocerebellar projections. I. Clim-
        bing fiber distribution in the vermis of cat cerebel-
        lum. J. Comp. Neurol. 174:417-488.
Hagiwara, S. (1983). "Membrane potential-dependent ion channels
        in cell membrane. Phylogenetic and developmental
        approaches." Raven Press, New-York.
Herrup, K., and Mullen, R.J. (1979). Regional variation and
        absence of large neurons in the cerebellum of the
        staggerer mouse. Brain Res. 172:1-12.
Hughes, A.F.W. (1968a). "Aspects of neural ontogeny." Academic
        Press, London.
Hughes, A.F.W. (1968b). Development of limb innervation, in :
        "Growth of the nervous system. A CIBA Foundation
        Symposium.", G.E.W. Wolstenholme and M. O'Connor, eds.
        J.A. Churchill, London, pp 110-117.
Innocenti, G.M., and Caminiti, R. (1980). Postnatal shaping of
        callosal connections from sensory areas. Exp. Brain Res.
        38:381-394.

Jan, Y.N., Jan, L.Y., and Dennis, M.J. (1977). Two mutations of synaptic transmission in Drosophila. Proc. R. Soc. B 198:87-108.

Koppel, H., and Innocenti, G.M. (1983). Is there a genuine exuberancy of callosal projections in development ? A qualitative electron microscopic study in the cat. Neurosci. Letters 41:33-40.

Kung, C., Chang, S.Y., Satow, Y., Van Houton, J., and Hansma H. (1975). Genetic dissection of behavior in Paramecium. Science 188:898-904.

Landis, D.M., and Sidman, R.L. (1978). Electron microscopic analysis of postnatal histogenesis in the cerebellar cortex of staggerer mutant mice. J. Comp. Neurol. 179:831-863.

Llinas, R., and Sugimori, M. (1979). Calcium conductances in Purkinje cell dendrites : their role in development and integration. In "Developmental and chemical specificity of neurones. Progress in Brain Research, Vol. 51." M. Cuénod, G.W. Kreutzberg and F.E. Bloom, eds., Elsevier, New-York, pp 323-334.

Llinas, R., and Sugimori, M. (1980a). Electrophysiological properties of in vitro Purkinje cell somata in mammalian cerebellar slices. J. Physiol. (Lond.) 305:171-195.

Llinas, R., and Sugimori, M. (1980b). Electrophysiological properties of in vitro Purkinje cell dendrites in mammalian cerebellar slices. J. Physiol. (Lond.) 305:197-213.

Mariani, J., and Changeux, J.P. (1980). Multiple innervation of Purkinje cells by climbing fibers in the cerebellum of the adult staggerer mutant mouse. J. Neurobiol. 11:41-50.

Mariani, J., and Changeux, J.P. (1981a). Ontogenesis of olivo-cerebellar relationship. I. Studies by intracellular recordings of the multiple innervation of Purkinje cells by climbing fibers in the developing rat cerebellum. J. Neurosci. 1:696-702.

Mariani, J., and Changeux, J.P. (1981b). Ontogenesis of olivo-cerebellar relationship. II. Spontaneous activity of the inferior olivary neurons and climbing fiber-mediated activity of cerebellar Purkinje cells in developing rats. J. Neurosci. 1:703-709.

Mariani, J., Crepel, F., Mikoshiba, K., Changeux, J.P., and Sotelo, C. (1977). Anatomical, physiological and biochemical studies of the cerebellum from reeler mutant mouse. Phil. Trans. B 281:1-28.

Meyer, R.L., and Sperry, R.W. (1976). Retinotectal specificity : chemoaffinity theory. In "Neural and behavorial specificity. Studies on the development of behavior and the nervous system, Vol. 3." Academic Press, New-York, pp 111-149.

Peacock, J.M., and Walker, C.R. (1983). Development of calcium
     action potentials in mouse hippocampal cell cultures.
     Dev. Brain Res. 8:39-52.
Purves, D. (1983). Modulation of neuronal competition by post-
     synaptic geometry in autonomic ganglia. Trends in
     Neurosci. 6:10-16.
Ramon y Cajal, S. (1911). "Histologie du Système Nerveux de
     l'Homme et des Vertébrés." Instituto Ramon y Cajal,
     Madrid.
Redfern, P.A. (1970). Neuromuscular transmission in new-born
     rats. J. Physiol. (Lond.) 209:701-709.
Salkoff, L., and Wyman, R. (1983). Ion channels in Drosophila
     muscle. Trends in Neurosci. 6:128-133.
Schwartzkroin, P.A. (1981). Development of rabbit hippocampus :
     physiology. Dev. Brain Res. 2:469-486.
Schwartzkroin, P.A., and Slansky, M. (1977). Probable calcium
     spikes in hippocampal neurons. Brain Res. 135:157-161.
Shimono, T., Nosaka, S., and Sasaki, K. (1976). Electrophysio-
     logical study on the postnatal development of neuronal
     mechanisms in the rat cerebellar cortex. Brain Res.
     108:279-294.
Sidman, R.L. (1972). Cell interactions in developing mammalian
     central nervous system. In "Cell interactions. Procee-
     dings of the Third Lepetit Colloqium." L.G. Silvestri,
     ed., North-Holland, Amsterdam, pp 1-13.
Sotelo, C. (1980). Mutant mice and the formation of cerebellar
     circuits. Trends in Neurosci. 3:33-36.
Sotelo, C., Bourrat, F., and Triller, A. (1984). Postnatal
     development of the inferior olivary complex in the
     rat. II. Topographic organization of the immature
     olivocerebellar projections. J. Comp. Neurol. in press.
Sotelo, C., and Changeux, J.P. (1974). Transsynaptic degenera-
     tion en cascade in the cerebellar cortex of staggerer
     mutant mice. Brain Res. 67:319-326.
Sperry, R.W. (1963). Chemoaffinity in the orderly growth of neu-
     ral circuits. Proc. Natl. Acad. Sci. (USA) 50:703-710.
Spitzer, N.C. (1982). The development of electrical excitabi-
     lity. In "Neuronal-glial cell interrelationships."
     T.A. Sears, ed., Springer-Verlag, Berlin, pp 77-91.
Thach, W.T. (1968). Discharge of Purkinje and cerebellar nuclear
     neurons during rapidly alterning arm movements in the
     monkey. J. Neurophysiol. 31:785-797.
Von der Malsburg, C., and Willshaw, D.J. (1977). How to label
     nerve cells so that they can interconnect in an ordered
     fashion. Proc. Natl. Acad. Sci. (USA) 74:5176-5178.
Woodward, D.J., Hoffer, B.J., and Altman, J. (1974). Physiolo-
     gical and pharmacological properties of Purkinje cells
     in rat cerebellum degranulated by postnatal X-irradia-
     tion. J. Neurobiol. 5:283-304.

Woodward, D.J., Hoffer, B.J., Siggins, G.R., and Bloom, F.E.
    (1971). The ontogenetic development of synaptic junctions,
    synaptic activation and responsiveness to neurotrans-
    mitter substances in the rat cerebellar Purkinje cells.
    Brain Res. 34:73-97.
Wu, C.F., and Ganetzky, B. (1980). Genetic alteration of nerve
    membrane excitability in temperature sensitive paraly-
    tic mutants of Drosophila melanogaster. Nature 286:
    814-816.

CNS HYPOMYELINATED MUTANT MICE (JIMPY, SHIVERER, QUAKING):

IN VITRO EVIDENCE FOR PRIMARY OLIGODENDROCYTE DEFECTS

Merrill K. Wolf and Susan Billings-Gagliardi

Department of Anatomy
University of Massachusetts Medical School
55 Lake Avenue North, Worcester, MA

Myelin is the protein and lipid sheath, consisting of consecutive, spirally wrapped, compacted segments of cell membrane, which surrounds each large axon in vertebrates and causes saltatory conduction of the action potentials. Nerve fiber sheaths with certain features of myelin are found in other animals, notably in some arthropods. True myelin, however, seems to be confined to true vertebrates, from the elasmobranchs up, and is strongly conserved in vertebrate evolution. Myelin has unique ultrastructural and biochemical components; it is formed by unique cells (in the CNS of warm-blooded vertebrates, by the oligodendrocytes); and it has a unique relationship to another cell, for normally it only forms around part of a neuron, almost always the axon. All this implies the existence of numerous DNA sequences which encode, not only for the structural components of adult myelin, but also for various steps of its development. Because myelin is strongly conserved, information about this DNA obtained from any mammal is likely to be directly applicable to other mammals including man.

. The single-gene mutations which impair CNS myelination in various mammals are therefore important "experiments of Nature" for the elucidation of the development and diseases of human CNS myelin. In the mouse, the mammal of choice for genetic studies, there are six such mutations presently known, all recessive, at four different genetic loci. One of these mutants, twitcher (twi), appears to produce a degenerative disease. Myelin forms normally, but then breaks down because the lack of the enzyme galactosylceramidase disrupts orderly lipid turnover. The other five mutations produce disorders of myelin development, and are the focus of our research interests. They are: jimpy (jp); myelin synthesis deficiency, which might appropriately be re-named mild

jimpy (jp^{msd}); quaking (qk); shiverer (shi); and myelin-
deficient, or, more appropriately, mild shiverer (shi^{mld}).

The histopathology of these mutant diseases has been reviewed
(Billings-Gagliardi and Wolf, 1982). Briefly, the sex-linked,
allelic jp and jp^{msd} mutations produce a dramatic deficiency of
mature oligodendrocytes and therefore of myelin, but each surviving
mature oligodendrocyte seems to produce a relatively normal quota
of myelin segments which are thin but otherwise ultrastructurally
normal. The two mutant diseases are identical except that jp^{msd}
has about twice as much myelin (and probably oligodendrocytes) as
jp (Wolf et al., 1983a). Mutant white matter tracts contain fat-
laden cells which look like fatty macrophages. However, since
there is no other morphological evidence of myelin destruction,
these cells may be profoundly abnormal derivatives of oligodendrocyte precursors. There is some evidence suggesting that oligodendrocyte precursors proliferate at normal or increased rates
(Skoff, 1982) but that most of them fail to express normal mature
biosynthetic activities (Bologa and Herschkowitz, personal communication). The numbers of myelin sheaths and of the fat-laden
cells vary systematically among various CNS white matter tracts,
both numbers being lower in late-myelinating tracts. PNS myelin
appears to be completely normal (Billings-Gagliardi and Adcock,
1981). The amount of CNS myelin may increase somewhat with age,
but the power of this observation is limited because affected jp
and jp^{msd} males do not survive much beyond weaning, i.e. 25 to
30 postnatal days.

The qk mutation, on chromosome #17, produces a substantial
reduction in the number of CNS myelin segments, but the number of
oligodendrocytes – at least, of cells which look like oligodendrocytes – is increased. Thus, it appears ipso facto that qk/qk
oligodendrocytes can not produce their normal quota of myelin
segments. There is suggestive evidence that qk myelin may have an
abnormal lipid composition (reviewed by Hogan, 1977). Its ultrastructure shows abnormalities and irregularities of wrapping. However, unlike the abnormalities of shi, which will be discussed,
those of qk myelin are non-specific and reversible. The lifespan
of qk/qk animals is at least several months, permitting the observation that the numbers of myelin sheaths and the orderliness of
their structure both increase with age, at least up to 120 postnatal days (Nagara and Suzuki, 1981). Peripheral myelin sheaths
in qk are normal in number and qualitative ultrastructure but are
too thin, having 30% to 40% fewer lamellae than normal for the
size of their axons.

The two allelic mutations, shi and shi^{mld}, have been more
recently identified. They both have a deficiency of a major
structural constituent of myelin, namely, myelin basic protein
(MBP). The severe allele, shi, has no detectable MBP in its CNS,

and lacks the major dense line completely at 21 postnatal days,
although small amounts of major dense line are seen in the CNS of
older (50 days) shi/shi animals. The mild allele, shi^{mld}, has
some MBP and some major dense line at all ages examined. In addition, both mutants show a number of striking and specific ultrastructural abnormalities: errors of compaction, wrapping, and targeting of myelin and sheaves of abnormal microprocesses of oligodendrocytes. These abnormalities, and the relationship between
the two alleles, will be described and discussed further below.
Surprisingly, although the PNS of these two mutants is also
severely deficient in MBP, PNS myelin is almost completely normal
in ultrastructure, including a normal-looking major dense line.

The goal of studying these mutations, and similar ones which
might be identified in the future, is to identify the defective
molecule specified by the mutant DNA and thereby to ascertain the
function of the normal molecule specified by the wild-type DNA
normally present at that locus. However, before a rational
approach to a mutant disease affecting a complex organism can be
designed at the molecular level, it is necessary to localize the
disease within the organism, both in space and in time. A defect
of myelin cannot be assumed, without proof, to be a defect of the
myelin-forming cell. Myelin might be rendered abnormal by the
presence of a systemic toxin, the lack of a systemic growth factor,
an autoimmune mechanism, or some other extrinsic cause. Even when
a myelin disease is shown to be intrinsic to the tissue containing
the myelin, it must then be asked which part of the tissue is defective: the myelin-_forming_ cell, the myelin-_receiving_ cell (i.e.,
the neuron), or some hypothetical "third element" of the tissue
which is not part of the final structure of the myelinated axon
but which, nevertheless, might affect the process of assembling
it. Furthermore, any mutant defect, irrespective of its cellular
target, might affect some early stage of development rather than
directly impairing an adult structural component. The final structure, although very abnormal, might be a remote epigenetic consequence of the genetic defect, not the direct expression of it.

The isolation of the myelin-forming tissues _in vitro_, separated
from the rest of the organism, provides a means to test such hypotheses. A culture system in which genetically normal tissue
makes myelin in the morphological as well as the chemical sense
provides, at once, a test of whether a given mutant affects the
myelin-forming tissue directly or through the rest of the organism.
Such cultures can potentially be further dissected in order to test
hypotheses about the cell type and stage of development involved.
Note the requirement that the cultures not only make the structural
components of myelin, but also assemble them into ultrastructurally
correct myelin sheaths. In the present state of the art, this
limits the choice of culture system. PNS tissue can be separated
in vitro into pure populations of neurons, Schwann cells, and fi-

broblasts. All three can be maintained in pure culture and reassembled at will to form normal myelin (Wood, 1976). Such experiments have demonstrated that all three cell types are essential (Bunge, Moya, and Bunge, 1981). CNS tissue is not yet similarly tractable. To produce CNS myelin _in vitro_ in the amount and morphological normalcy necessary to validate a mutant study, one must accept a relatively untidy system: the complex neuropil generated by primary explants, so-called "organotypic" cultures, maintained on an undefined medium using such biological products as serum, embryo extract, and reconstituted collagen gel. However, the complexity of the system, if it is understood, can not merely be tolerated, but actually used to assist the experimenter's purposes. Our studies have employed cultures of genetically normal postnatal day 0 (P-0) mouse cerebellum. Such cultures produce myelin in the abundance needed for mutant studies. In addition, the principal types of neurons of cerebellar cortex and deep nuclei develop much of their specialized form and synaptic relationships _in vitro_. This provides an important control for the specificity of the effect of the mutant genes, all of which affect myelin, but not neurons and their synaptic relationships. While these cultures cannot be reduced to individual cell types, it is now possible to remove and add back the myelin-forming oligodendrocytes in controlled experiments.

We began our culture studies with the mutants jp, jp^{msd}, and qk; those studies have been reviewed (Billings-Gagliardi and Wolf, 1982; Wolf and Billing-Gagliardi, 1982; Billings-Gagliardi et al., 1983). In brief, all three of these mutant diseases are reproduced _in vitro_ in the following senses: Cultures are very deficient in myelin, but completely normal in development of neuron somas, dendrites, and synaptic interrelationships. Quantitative comparison of cultures of the three mutants and their controls shows that the degree of myelin deficiency _in vitro_ is in keeping with the quantitative deficiency of each mutant _in situ_. The myelin of jp and jp^{msd} appears, _in vitro_ as _in situ_, to be elaborated by a small number of oligodendrocytes which escape the general consequence of the genetic defect, each one producing a relatively normal cluster of myelin segments. The evidence for abnormal lipid composition of qk/qk myelin receives odd corroboration from the finding that living qk/qk myelin sheaths in the cultures are invisible by a light microscopic technique which demonstrates not only control myelin but also the much smaller number of myelin sheaths found in living jp and jp^{msd} cultures. Since the "invisibility" of living qk/qk myelin is not correlated with lack of compaction or with any other ultrastructural feature, it must reflect an abnormal refractive index of the living cell membrane.

The deficiency of jp myelin at first appeared to be relatively greater _in vitro_ than _in situ_. However, those initial studies used the available Jackson Laboratory stock of jp mice, which differed

in the remainder of the genome from the jp^{msd} stock. Since the
jp^{msd} stock was derived from two inbred strains both of which
were known to be good for our culture purposes, while the jp stock
contained genes from a different inbred strain which was untested
in this regard, we transferred the jp gene to a stock strictly comparable to the jp^{msd} stock (Wolf et al., 1983a). On the new
background, the relationship between jp and jp^{msd} in situ remained the same as before. However, the myelination of jp cultures
was improved enough to make the relationship between jp and jp^{msd}
myelination in vitro the same as the relationship in situ. This
demonstrated the importance of deriving all our cultures of mutant
brain from stocks which would be strictly comparable in all of the
genome other than the mutant loci.

The severed optic nerve provides a tract of CNS white matter
free of neuron cell bodies and therefore of viable axons. (Other
tracts, such as the corpus callosum, could theoretically be used,
but are impossible to dissect free of neuron somas from an immature
mouse brain.) When a fragment of genetically normal P-0 to P-14
optic nerve is added to a culture of cerebellum from any of the
three mutants, with the cut surfaces of the optic nerve in direct
contact with the cerebellum (a procedure which we call "glial injection"), the myelin content of the cerebellum is dramatically
increased in the immediate vicinity of the zone of fusion between
the piece of nerve and the cerebellum, and to a varying extent in
adjacent regions of the cerebellum explant. The strict requirement for a zone of fusion, and other positional, quantitative, and
statistical evidence, strongly suggest that the additional myelin
is directly made by normal oligodendrocytes myelinating mutant
axons. Thus, the mutant axons are competent to evoke and accept
myelination by normal oligodendrocytes; no other element present
in the mutant tissue interferes with the process; and the defect
in each of these three mutants can be presumptively assigned to
the oligodendrocyte. The argument that the additional myelin is
made by the injected normal glia is strengthened by evidence from
radioactive labelling experiments (Billings-Gagliardi et al.,
1983). Cultures of non-radioactive jp^{msd} cerebellum were injected after 4 days in vitro, with optic nerve from P-14 mice
which had received 6 uc of 3H-thymidine every 12 hours for 3 days
prior to sacrifice. After 14 additional days in vitro the cultures
were fixed and sectioned. Heavily labeled oligodendrocytes were
identified immediately adjacent to the zone of fusion, and more
lightly labeled ones farther from the nerve, suggesting that oligodendrocytes from the donor nerve not only invaded the host tissue,
but diluted out their label by cell division as they migrated. Two
labeled cells, identified in close apposition to myelinated axons,
were proved to be oligodendrocytes by subsequent re-embedding and
transmission electron microscopy of the same semithin section initially used for autoradiography.

The key limitation of these experiments so far is that it is
not possible to identify the genotype of any given oligodendrocyte
or myelin sheath in the injected cultures. The few oligodendrocytes in jp and jp^{msd} brains or cultures, and the myelin segments
they make, look normal. Their relative thinness in situ would not
be recognizable in vitro because the quantitative relationship between axon size and thickness of myelin in normal CNS is not reproduced by the cultures. The abnormalities recognizable on ultrastructural examination of qk myelin are completely non-specific.
They differ only in degree from the non-specific irregularities
seen in all CNS myelin in vitro, even in the best cultures. In
the one autoradiographic experiment, physical continuity of a
labeled oligodendrocyte and a myelin sheath was not demonstrated.
Indeed, the present state of electron microscopic technique does
not permit this continuity to be demonstrated at will in any speci-
men of interest. Thus, all the evidence collected up to this point
to support our interpretation of the glial injection experiment
could be criticized as circumstantial, not absolute. Unfortu-
nately, the immunological and genetic markers so far developed to
distinguish cells from mice of different inbred strains do not work
on oligodendrocytes or on myelin. Another problem at this point
was that the reciprocal experiment, injecting mutant optic nerve
into a culture of normal cerebellum, was unlikely to be interesting
in this simple form. The normal oligodendrocytes in the cultures
would simply swamp out the abnormal ones. To address these problems, we turned to an additional, more recently discovered mutant,
and to a modification of the culture method.

The two mutants at the shiverer locus, as already stated, are
deficient in MBP. The severe allele, shi, has no detectable MBP
in its CNS and has the following specific and absolute morphological defects. Oligodendrocytes produce sheaves of prodigiously
numerous, abnormal microprocesses. (Uninitiated observers might
confuse these bundles with bundles of unmyelinated axons, but the
processes are much smaller in diameter and more electron-dense.)
Myelin is often inappropriately applied to these bundles or to
oligodendrocyte perikarya, (Fig. 1). Myelin is often incorrectly
wrapped, not in closed circular spirals but in open stacks which
superficially resemble compressed sine-waves. Myelin is usually
not compacted; that is, oligodendrocyte cytoplasm is usually retained between the cytoplasmic surfaces of the oligodendrocyte
membranes, although the external surfaces are apposed to form an
intraperiod line (Figs. 2-3). For reasons not yet understood, the
intraperiod line is denser in shi/shi than in normal myelin. Even
when there is "pseudo-compaction," with removal of the cytoplasm
from between the cytoplasmic faces of the membrane, a normal major
dense line is categorically never observed in 20-day old shi/shi
CNS myelin. At 50 postnatal days and later, some bits of true
major dense line may be observed, but the other abnormalities
persist unabated in severity to the oldest ages we have examined,

120

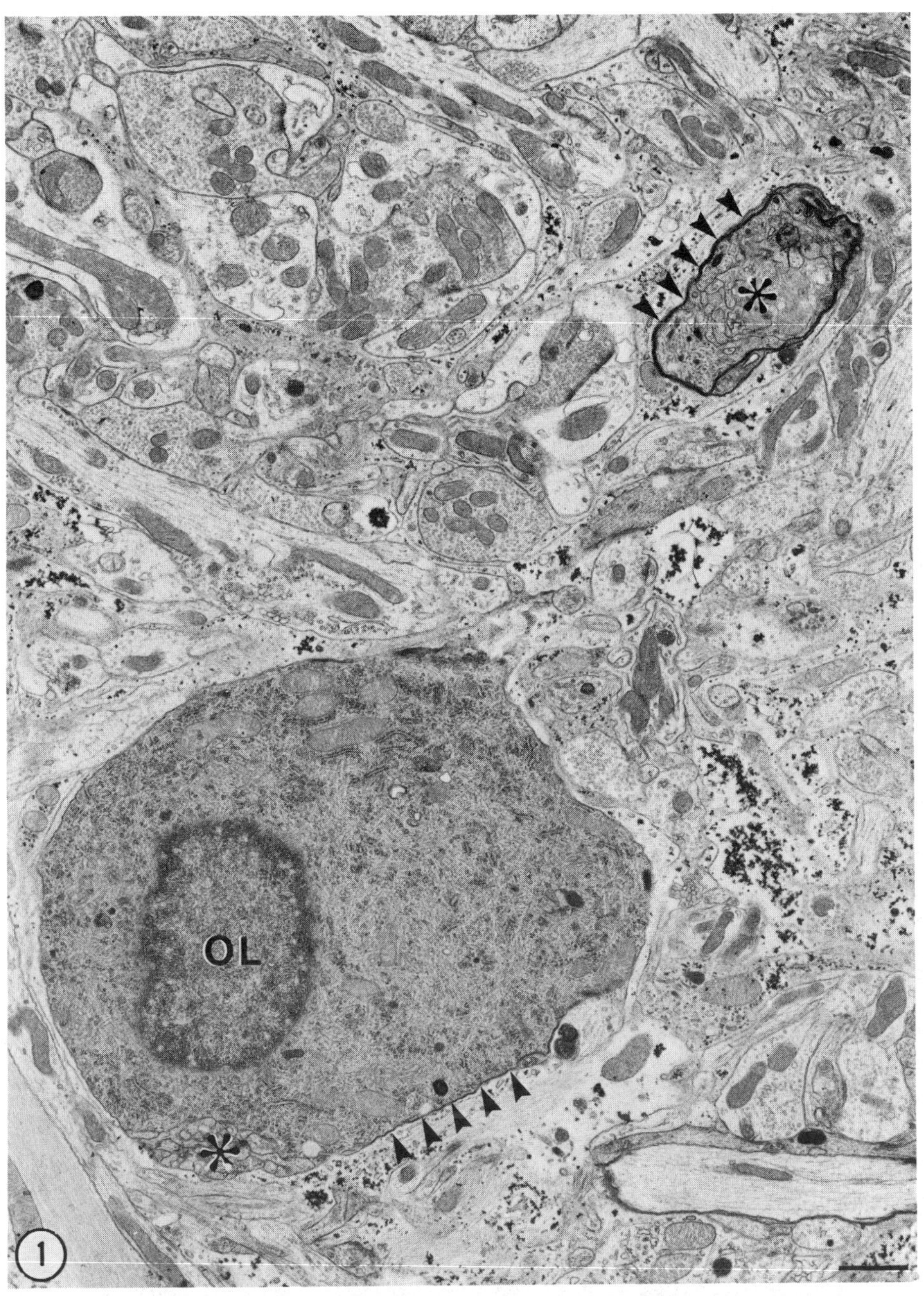

Fig. 1. <u>shi</u> cerebellum <u>in vitro</u>. As <u>in situ</u>, oligodendrocytes (OL) have microprocesses arranged in bundles (*). These bundles, as well as the oligodendrocyte cell body, may be targets for myelination (arrowheads). Bar = 1.0 um.

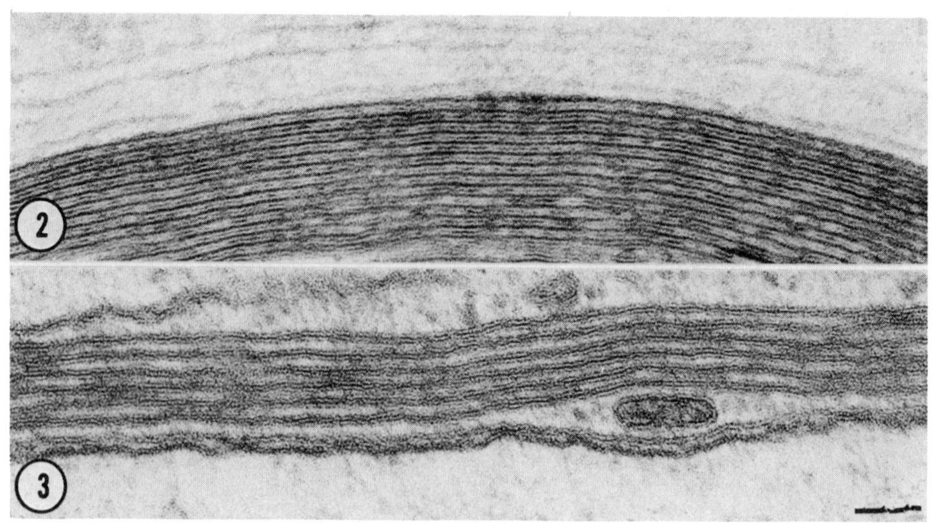

Fig. 2.   Myelin in a normal cerebellar culture at 20 days _in vitro_.
Fig. 3.   Myelin in a _shi_ culture at 20 days _in vitro_. In contrast
          to normal myelin in culture (or _in situ_), _shi_ myelin has
          no major dense line even where cytoplasmic faces of the
          membrane are apposed. In addition, the density of the
          intraperiod line is increased relative to normal myelin.
          Bar = 0.1 um. (From Billings-Gagliardi et al., in press.)

namely, 111 postnatal days. As an essential prelude to our culture
work, we used a standard cross-intercross breeding scheme to transfer the _shi_ mutation to a hybrid background comparable to that of
_jp_ and _jp_^{msd}. On our hybrid background, _shi/shi_ CNS seems to
produce more myelin than on other backgrounds, and this facilitates the assembly of a representative sample of the abnormalities
described above. The abnormalities, however, are unaltered on the
new background. Two other circumstances have facilitated the work.
Affected _shi/shi_ males and females are both fertile and can produce
litters guaranteed to consist entirely of affected offspring: this
circumvents the problems presented by the other mutants of using
"marker genes" to determine which newborn animals express the
mutant disease. Secondly, MBP, like many other proteins, can be
demonstrated in tissues by immunocytochemical techniques, but its
affinity for its antibody is exceptionally hardy. Immunocytochemical localization of most proteins necessitates compromises in
tissue preservation because of the minimal fixation, frozen or
Vibratome sectioning, and other special methods required. MBP,
however, can be demonstrated by this means in tissue correctly
fixed and embedded in plastic for transmission electron microscopy, by deplasticizing the sections and using a standard
peroxidase-antiperoxidase procedure (Trapp et al., 1981).

The defects of shi/shi are reproduced in organotypic cultures.
However, because of the categorical, specific nature of the shi/shi
abnormalities, their recognition in vitro does not depend on quantitative, distributional, or other circumstantial factors; the defects can be recognized at an individual level. Myelin is somewhat more abundant in cultures of shi/shi than in jp, jp^{msd}, or
qk cultures, in keeping with the relative abundance of myelin observed in situ in our stock of shi/shi mice. When this myelin is
compared with a genetically normal control culture, virtually every
piece of myelin seen can be assigned to the correct genotype on the
basis of its individual ultrastructure, without needing to refer to
the context of the surrounding tissue. The abnormal oligodendrocyte microprocesses are also observed in shi/shi cultures, where
they are, if anything, even more exuberant and disorderly than in
situ. We speculate that the relative looseness of culture neuropil
frees the processes from spatial contsraints present in white
matter in situ.

Immunocytochemistry shows that normal control cultures contain
numerous MBP-positve myelin profiles. Cultures of shi/shi contain
numerous myelin profiles, but all of them are MBP-negative and
invisible in the immunocytochemical preparations. When shi/shi
cultures injected with normal optic nerve are examined, MBP-
positive degenerating myelin is seen within the optic nerve profile, and MBP-positive intact myelin sheaths are seen in the cerebellum near the zone of fusion and in the surrounding part of the
culture (Figs 4-6).

These findings are confirmed by transmission electron microscopy on the same series of injected cultures, which shows that
such cultures contain two ultrastructurally specific kinds of
myelin, normal and shi/shi (Figs 7-8). Most of the culture contains only shi/shi myelin, but as the optic nerve is approached,
an increasing number of normal myelin profiles occur. Near the
zone of fusion with optic nerve, myelin segments are more numerous
than elsewhere in the culture. Most of the myelin in this part of
the culture is normal, but normal and shi/shi myelin segments can
occur within a few micra of each other. We have seen normal and
shi/shi myelin segments adjacent to each other on the same axon,
separated only by a single standard node of Ranvier. Thus, shi/shi
and normal oligodendrocytes each make myelin with their own characteristic ultrastructure and protein content around shi/shi axons
(Billings-Gagliardi et al., in press).

How would oligodendrocytes of these two genotypes behave in a
similar comparative confrontation with normal axons? This was approached by treating cultures with the "false nucleotide" cytosine
arabinoside (Ara-C), which is preferentially toxic to dividing
cells. In cultures of PNS, the schedule of Ara-C treatment can be
controlled so as to destroy either all the non-neuronal cells or

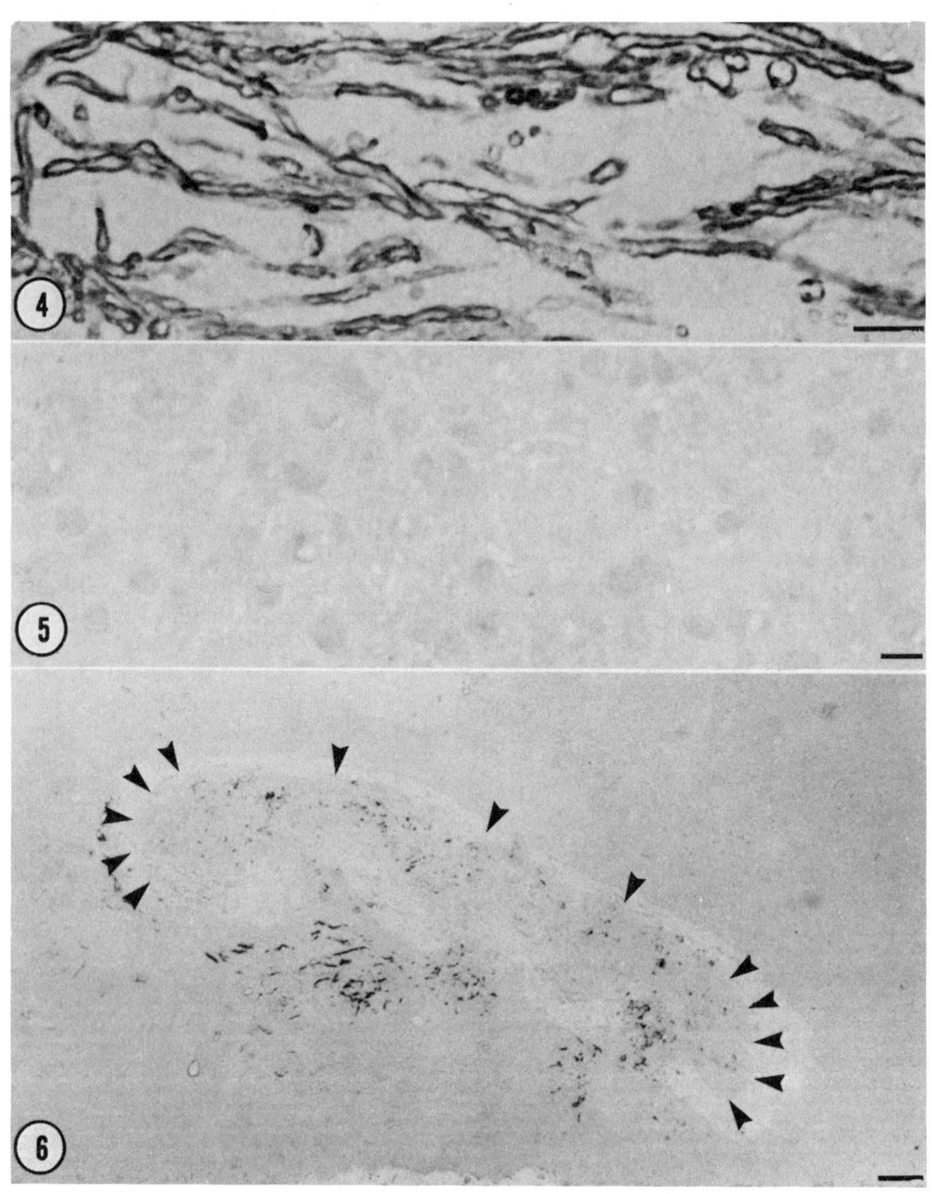

Figs. 4-6. Cultures stained immunocytochemically for myelin basic protein (MBP). Figs. 4-5, bars = 10 um; Fig. 6, bar = 50 um. (From Billings-Gagliardi et al., in press.)

Fig. 4. Normal culture: abundant MBP+ myelin.

Fig. 5. <u>shi</u> culture: no MBP staining, although presence of myelin in adjacent sections confirmed by conventional staining.

Fig. 6. <u>shi</u> culture + normal optic nerve: MBP+ myelin near the optic nerve (outlined by arrowheads).

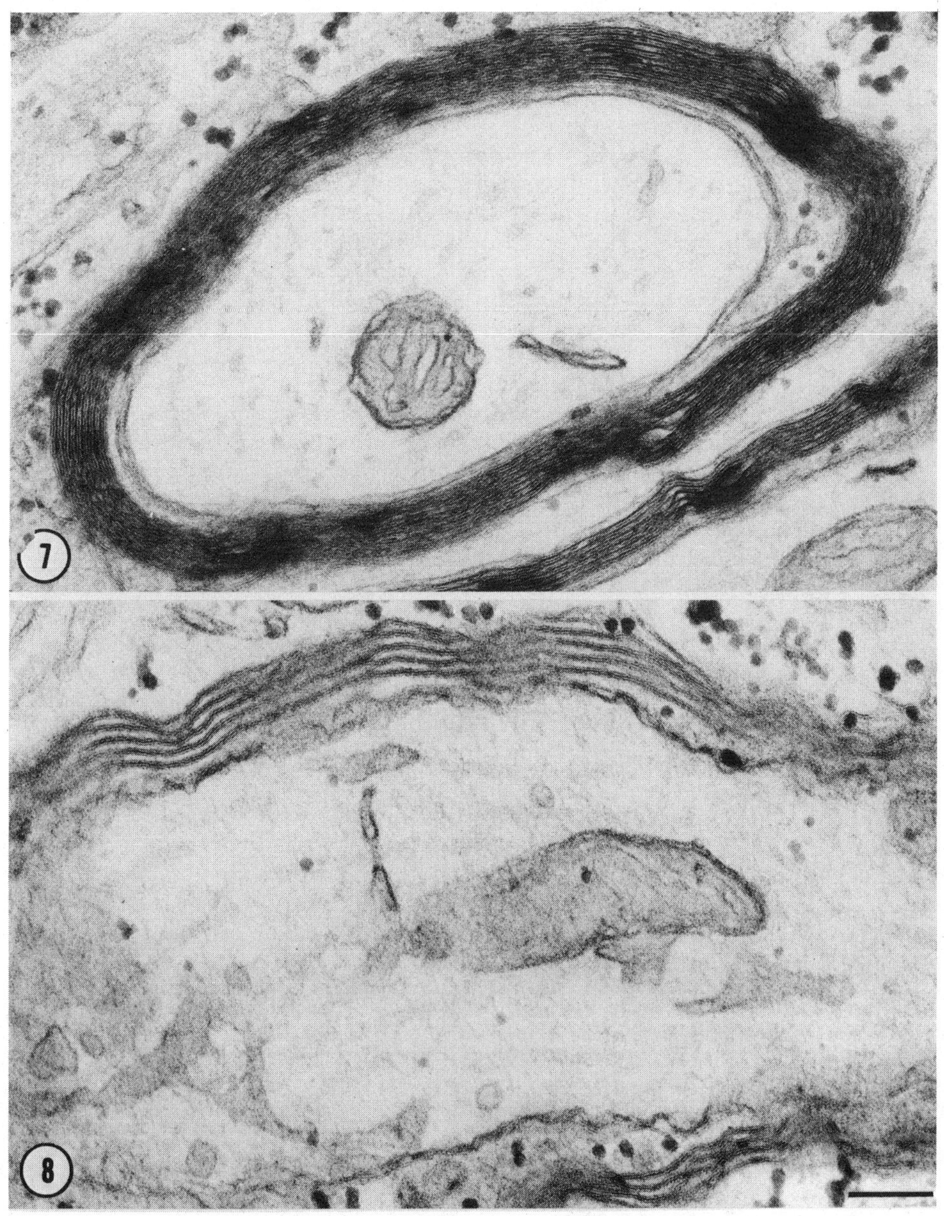

Figs. 7-8. Adjacent ultrathin sections of the injected culture in
Fig. 6. In this culture _shi_ axons near the optic nerve
may be surrounded by either ultrastructurally normal
(Fig. 7) or _shi_ (Fig. 8) myelin. At a distance from
the optic nerve, only _shi_ myelin is present. This
mirrors the distribution of MBP+ myelin. Bar = 0.2 um.
(From Billings-Gagliardi et al., in press.)

only the fibroblasts, which are more sensitive to the drug than
Schwann cells (Wood, 1976). In cultures of cerebellum, Ara-C was
originally used by Seil (Seil et al., 1980) to destroy the granule
cell neurons. Only later (Seil and Blank, 1981) did he emphasize
concomitant destruction of myelinating oligodendrocytes. We found
that in cultures of cerebellum treated with 5 ug/ml Ara-C for the
7 or 8 days in vitro, then washed free of Ara-C and maintained for
an additional 12 days, myelin and recognizable oligodendrocytes
were totally absent (Figs. 9-10). The cultures contained numerous
healthy neuron somas bearing large diameter axons which normally
would have been myelinated and presumably were competent to accept
myelination by oligodendrocytes re-introduced into the cultures.

Our cultures proved somewhat more resistant to Ara-C than
Seil's cultures, no doubt reflecting numerous known differences
both in minor details of our culture method and in the genotype of
our normal mice. We needed 7 or 8 days of Ara-C treatment; the 5
day treatment used by Seil was not enough. Granule cells were reduced in numbers but never totally eliminated from our cultures.
The observation of mitotic figures in the cultures suggests that
some other mitotically competent cells may also survive. Astrocytes not only survive but show hypertrophy and probably are also
increased in numbers, although counts have not yet been done.
Thus, in our hands, Ara-C treatment provides a method to selectively remove the myelinating oligodendrocytes from cultures of
cerebellum with relatively less damage to every other cell type
(Schwing-Stanhope and Wolf, 1982; Wolf et al., 1983b).

When oligodendrocytes are re-introduced into such cultures, by
injecting them with optic nerve after completing the Ara-C treatment and removing the drug, the cultures produce myelin (Fig. 11).
As in cultures of myelin-deficient mutant cerebellum injected with
normal optic nerve, the myelin in these cultures is concentrated
around the fusion zone between cerebellum and optic nerve. We
have the impression that oligodendrocytes migrate further or more
rapidly in Ara-C treated normal cultures than in untreated cultures
of mutant cerebellum. Perhaps the killing of many cells by the
Ara-C opens up channels in the tissue and facilitates the migration. The phenotype of the myelin depends entirely on the source
of the optic nerve. Normal optic nerve produces only normal
myelin, while shi/shi optic nerve produces only shi/shi myelin
(Wolf et al., 1983c) (Figs 12-14). The properties of the shi/shi
oligodendrocytes are not altered either by the genetically normal
axons they are myelinating or by the numerous genetically normal
astrocytes present in the surrounding neuropil.

Thus, normal oligodendrocytes make normal myelin, and shi/shi
oligodendrocytes make shi/shi myelin, irrespective of whether they
are myelinating normal axons in a normal tissue environment or
shi/shi axons in a shi/shi tissue environment. This is nearly

126

conclusive evidence against the involvement of the axon or of a
"third element" in the shi defect. Unless some strong positive
evidence to the contrary is obtained in the future, the shi mutations can be considered to produce an intrinsic defect in the
oligodendrocyte. Furthermore, this proves that in glial injection
experiments involving shi/shi and normal tissues the introduced
oligodendrocytes make myelin themselves, rather than exerting some
sort of humoral influence upon the resident oligodendrocytes. This

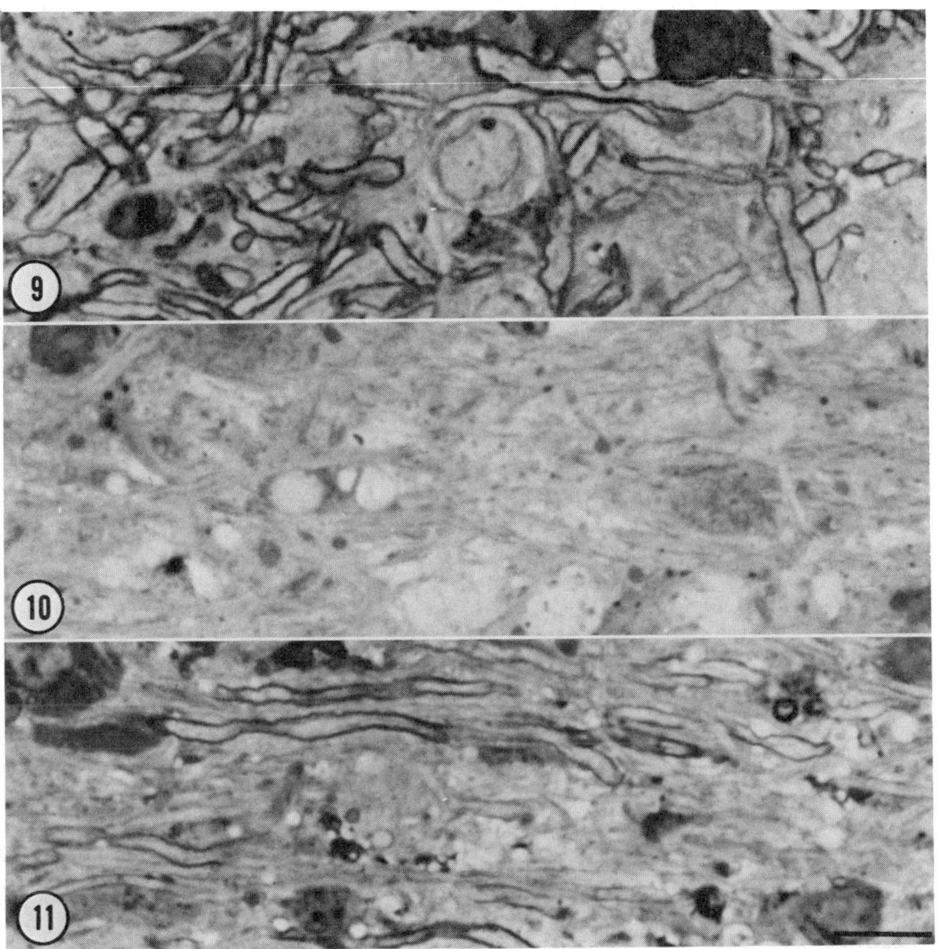

Figs. 9-11. Explants of cerebellum from the same mouse at 20
     days in vitro. Bar = 10 um.
Fig. 9. Untreated: Abundant myelin.
Fig. 10. Treated with Ara-C (days 1-7 in vitro): No myelin or
          mature oligodendrocytes.
Fig. 11. Treated with Ara-C then injected with normal optic nerve:
          Abundant myelin near injected normal optic nerve.

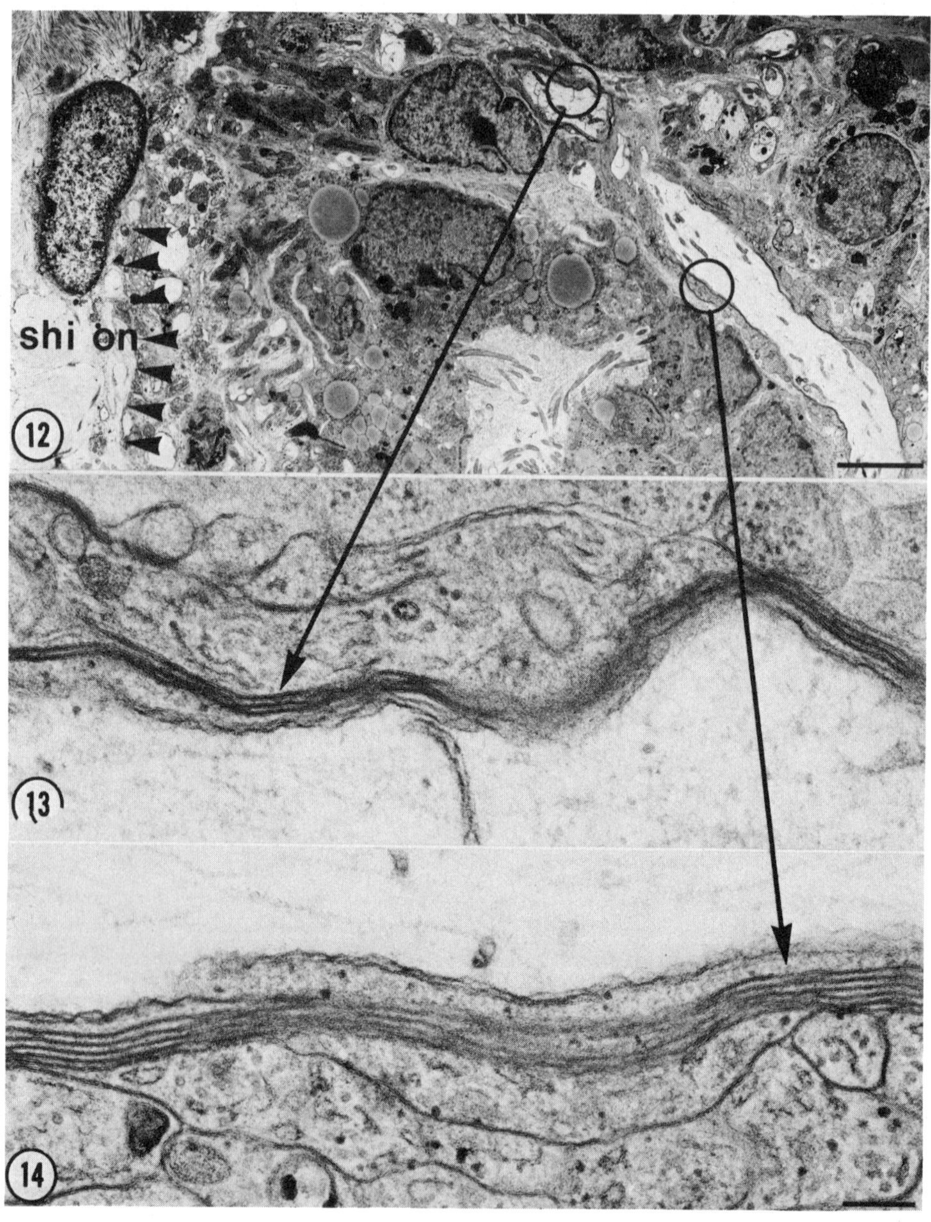

Figs. 12-14. Normal cerebellar explant treated with Ara-C, then injected with <u>shi</u> optic nerve.

Fig. 12. Culture has numerous myelinated axons near the injected optic nerve (on) at 20 days <u>in vitro</u>. Bar = 5 um.

Figs. 13-14. <u>All</u> the myelin in the culture is of the <u>shi</u> type, uncompact and lacking the major dense line. Bar = 0.2 um.

validates our interpretation of the source of the additional normal
myelin in glial injection experiments with the other mutants, and
thus supports our arguments, based on those experiments, that these
mutants also have intrinsic defects of their oligodendrocytes. The
white matter in $\underline{jp}$ and $\underline{jp^{msd}}$ animals shows a marked astrocytosis.
Skoff (1976) has argued that an astrocytic abnormality may con-
tribute to the myelin deficiency in these mutants, although his interpretation has been challenged (Meier and Bischoff, 1977; Privat
et al., 1979). We see a moderate astrocytosis in $\underline{shi/shi}$ white
matter as well. The experiments discussed above make it almost
certain that this astrocytosis is merely the normal reaction of
this cell type to an abnormality of the surrounding tissue, i.e.,
the myelin and oligodendrocytes. We consider that the more promi-
nent astrocytosis seen in $\underline{jp}$ and $\underline{jp^{msd}}$ brains is also the normal
reaction of this cell type to the even greater deficiency of myelin
in these mutants.

It would appear that the myelin-deficient mutations are
leaky. Myelinating oligodendrocytes are not totally absent
from $\underline{jp}$ CNS, nor is MBP totally absent from that of $\underline{shi/shi}$.
As $\underline{shi/shi}$ brain ages, it begins to acquire some bits of true
major dense line. The mechanism of the leak may be absolutely
crucial to understanding the mechanism of the mutation. Therefore, the allelic mutations with a milder defect, or, to put
it another way, a larger leak, deserve as much study as their more
severe alleles. We have preliminary data on the mild shiverer,
$\underline{shi^{mld}}$ (Shen et al., 1983). Affected brain looks exactly like
$\underline{shi/shi}$, including the abnormal oligodendrocyte processes and the
errors of targeting, wrapping, and compaction of myelin, except for
a small minority of the myelin segments. While most myelin segments look just like those of $\underline{shi/shi}$, some of them have a substantial amount of true compaction and true major dense line. A very
small number, probably less than 5% of all the myelin segments, are
correctly wrapped and completely compacted, with a continuous major
dense line, and could be confused with genetically normal myelin.
The defects of $\underline{shi^{mld}}/\underline{shi^{mld}}$ are all completely reproduced in
culture, like those of $\underline{shi/shi}$, and injection of normal optic nerve
produces much additional normal myelin near the site of injection.
Note, however, that because of the few normal-looking sheaths pro-
duced by $\underline{shi^{mld}}/\underline{shi^{mld}}$, it is not possible to assign a genotype
to any individual segment of normal myelin in an injected culture.
Immunocytochemical evidence is extremely preliminary, and we do not
yet have positive correlations with ultrastructure. The brains of
$\underline{shi^{mld}}/\underline{shi^{mld}}$, and cultures derived from them, both contain
rare MBP-positive myelin segments. To our surprise, these are not
randomly distributed, but are clustered, as if a few oligodendrocytes somehow escape the genetic defect and make most of the MBP in
this mutant. Richard Mullen has suggested (personal communication)
that these cells may be somatic back-mutations to wild type. We
hope it may become possible to test this interesting idea.

Another, completely unexpected, piece of evidence about the leak in shi/shi has been provided by double mutant animals, which are both shi/shi and jp/Y. Carson and Kerner have prepared such animals (personal communication), by three consecutive generations of crossing jp carrier females to shi/shi males, and have found that the two mutant diseases tend to "cure" each other in the following sense. Brain of shi/shi lacks virtually all MBP, but has a substantial though not normal amount of myelin proteolipid protein. Brain of jp/Y has no detectable proteolipid protein, in keeping with the very small amount of myelin, but has more MBP than expected. Double mutant brains, which should be totally deficient in both proteins if each disease were expressed unmodified by the other, instead have measurable levels of both. We are preparing double mutants, using the mild as well as the severe alleles at both loci. Extremely preliminary results suggest that double mutant myelin looks like shi/shi except that there is more dense line than shi/shi alone should have at that age. We expect that more complete observations of these animals will contribute to detailed hypotheses about the mutants at both the shi and jp loci.

Finally, it is noteworthy that the increased amount of MBP introduced into the CNS of mutants at the shi locus by enlarging the leak does nothing to correct the cytological defects, other than the lack of the major dense line. They are not progressively corrected in shi/shi brains or in shi^{mld}/shi^{mld} brains as they age, nor are they less severe in the leakier mutant. It should also be noted that these other defects resemble, in exaggerated form, the behavior of normal immature oligodendrocytes. When these cells begin to make myelin, they make numerous mistakes of targeting and wrapping. Normally, however, the mistakes are corrected before many lamellae of myelin are laid down in erroneous patterns. Friedrich and Sternberger (1983) have immunocytochemical evidence that immature normal oligodendrocytes have numerous microprocesses which are presumably retracted as the cells mature. Are shi and shi^{mld} oligodendrocytes arrested at an immature stage of their development? If so, either the amount of MBP provided by the forms of the leak which have been observed so far is not enough to permit the cells to mature, or the immaturity and the shortage of MBP are two separate effects of the same mutation. We have already noted that the final abnormal structure produced by any mutation may be a remote epigenetic consequence of the primary mutant defect. We close with the suggestion that this must be considered even in a mutation like shi which seems to produce a straightforward defect in a single well-characterized protein.

Supported in part by NIH Grant NS-11425

130

REFERENCES

Billings-Gagliardi, S., and Adcock, L. H., 1981, Hypomyelinated
     mutant mice IV: PNS myelin of jp^{msd} is normal in quantity
     and ultrastructure, Brain Res., 225:309.
Billings-Gagliardi, S., and Wolf, M. K., 1982, CNS hypomyelinated
     mutant mice: Morphological and tissue culture studies, in:
     "Advances in Cellular Neurobiology," Vol. 3, S. Fedoroff and
     L. Hertz, ed., Academic Press, New York, 275.
Billings-Gagliardi, S., Adcock, L. H., Lamperti, E., Schwing-
     Stanhope, G. and Wolf, M. K., 1983, Myelination of jp
     jp^{msd}, and qk axons by normal glia in vitro: Ultra-
     structural and autoradiographic evidence, Brain Res., 268:
     255.
Billings-Gagliardi, S., Hall, A. L., Stanhope, G. B., Altschuler,
     R. J., Sidman, R. L., and Wolf, M. K., Cultures of shiverer
     mutant cerebellum injected with normal oligodendrocytes make
     both normal and shiverer myelin, Proc. Natl. Acad. Sci.
     USA, in press.
Bologa, L., Moll, C., and Herschkowitz, N., 1983, Normal prolifera-
     tion rate of galactocerebroside positive oligodendrocytes in
     brain cell cultures of the hypomyelinated mouse mutant
     jimpy, Brain Res., 275:369.
Bunge, R. P., Moya, F., and Bunge, M. B., 1981, Observations on the
     role of Schwann cell secretion in Schwann-cell-axon inter-
     action, in: "Neurosecretion and Brain Peptides", J. B.
     Martin, S. Reichlin, and K. L. Bick, eds., Raven Press, New
     York, 229.
Friedrich, V. L., and Sternberger, N. H., 1983, The "lacy" oligo-
     dendrocyte: an immature form revealed by immunocytochemical
     staining, Anat. Rec., 205:58A.
Hogan, E. L., 1977, Animal models of genetic disorders of myelin,
     in: "Myelin," P. Morell, ed., Plenum Press, New York and
     London, 489.
Meier, C., and Bischoff, A., 1977, Dysmyelination in jimpy mouse
     due to astroglial hyperplasia? Nature (Lond.), 268:177.

Nagara, H., and Suzuki, K., 1981, Chronological study of oligo-
     dendroglial alterations and myelination in quaking mice,
     Neuropath. Appl. Neurobiol., 7:135.
Privat, A., Drian, M. J., and Escaig, J., 1979, Jimpy mouse myelin
     revisited with freeze-fracture, Acta Neuropath., 45:129.
Schwing-Stanhope, G., and Wolf, M. K., 1982, Myelination of cyto-
     sine arabinoside-treated cerebellum cultures by oligodendro-
     cytes from normal optic nerve, Soc. Neurosci. Abst., 8:231.
Seil, F. J., Leiman, A. L., and Woodward, W. R., 1980, Cytosine
     arabinoside effects on developing cerebellum in tissue
     culture, Brain Res., 186:393.
Seil, F. J., and Blank, N. K., 1981, Myelination of central nervous
     system axons in tissue culture by transplanted oligodendro-
     cytes, Science, 212:1407
Shen, X-Y., Hall, A. L., Wolf, M. K., and Billings-Gagliardi, S.,
     1983, Myelin-deficient mouse mutation: comparison with
     allelic shiverer mutation in situ and in vitro, Soc.
     Neurosci. Abst., 9:6.
Skoff, R. P., 1976, Myelin deficit in the jimpy mouse may be due to
     cellular abnormalities in astroglia, Nature (Lond.), 264:
     560.
Skoff, R. P., 1982, Increased proliferation of oligodendrocytes in
     the hypomyelinated mouse mutant - jimpy, Brain Res., 248:
     19.
Trapp, B. D., Itoyama, Y., Sternberger, N. H., Quarles, R. H., and
     Webster, H. deF., 1981, Immunocytochemical localization of
     P_O protein in Golgi complex membranes and myelin of
     developing rat Schwann cells, J. Cell Biol., 90:1.
Wolf, M. K., and Billings-Gagliardi, S., 1982, A tissue culture
     strategy for studying mutant mice with CNS hypomyelination,
     in: "Neuroscience Approached through Cell Culture," Vol.
     II, S. Pfeiffer, ed., CRC Press, Boca Raton, 141.
Wolf, M. K., Kardon, G. B., Adcock, L. H., and Billings-Gagliardi,

S., 1983a, Hypomyelinated mutant mice V: Relationship
between jp<u>msd</u> and jp re-examined on identical genetic
backgrounds, <u>Brain Res.</u>, 271:121.

Wolf, M. K., Schwing-Stanhope, G., Hall, A. L., and Billings-
Gagliardi, S., 1983b, Cytosine arabinoside selectively
eliminates myelinating oligodendrocytes from cultures of
cerebellum, <u>Anat. Rec.</u>, 205:218A.

Wolf, M. K., Schwing-Stanhope, G., Hall, A. L., and Billings-
Gagliardi, S., 1983c, Shiverer oligodencrocytes form
shiverer myelin around normal axons, <u>Soc. Neurosci. Abst.</u> 9:
6.

Wood, P. M., 1976, Separation of functional Schwann cells and
neurons from normal peripheral nerve tissue, <u>Brain Res.</u>,
115:351.

EFFECTS OF ALTERATIONS OF CELL SIZE AND NUMBER ON THE STRUCTURE AND
FUNCTION OF THE XENOPUS LAEVIS NERVOUS SYSTEM

R. Tompkins, B. Szaro, D. Reinschmidt, C. Kaye and C. Ide

Department of Biology
Tulane University
New Orleans, LA

INTRODUCTION

The evolution of the vertebrate central nervous system has involved increasing cell numbers and overall relative size, allometric
changes in different brain regions, and quantitative and qualitative
alterations in the interactions between different regions of the
nervous system. The importance of genetic changes which alter the
metrics of interactions between developing organ systems for evolutionary change has been stressed by Gould (1977). Yet relatively
little evidence is available from experimental studies to support
the idea that allometric changes, such as changes in cell size, cell
number or the relative sizes of different organs within the vertebrate nervous system can cause qualitative as well as quantitative
changes in structure and function. Recent advances in the genetics
of Xenopus laevis have permitted us to investigate some of these
effects. The results support the hypothesis that relatively simple
alterations in cell size and numbers can cause both qualitative and
quantitative changes in morphology and function of the nervous
system.

Xenopus laevis is a member of a group of frogs which have had a
slow rate of morphological evolution. Speciation has in many in-
stances been associated with polyploidization and Xenopus laevis
probably arose as an allotetraploid about 30 million years ago
(Kobel, 1981). This mechanism of speciation, relatively common in
lower vertebrates such as amphibians (Bogart, 1980), causes changes
in cell size and cell numbers in all organs. It has proven relatively easy to construct polyploid amphibians and a number of strategies
for producing both autopolyploids and allopolyploids have been devised (Fankhauser, 1955). We have recently constructed autotetra-

ploid <u>Xenopus laevis</u> by pressure suppression of first cleavage in diploid zygotes (Reinschmidt et al., 1979) and from these have established a fertile tetraploid line which has been chromosomally stable for three generations.

In contrast to many invertebrates, polyploid vertebrates attain approximately the same size as their diploid progenitors (Fankhauser, 1955). Experimentally produced tetraploid animals retain the same nucleocytoplasmic ratio as diploids and hence each organ is built of about one-half the number of cells as are found in the comparable organs of diploid animals (Fankhauser, 1955). Although polyploidy initially results in an increase in the number of copies of each gene, diploidization of most loci occurs over a protracted period of time subsequent to speciation by polyploidization (Ohno, 1970). Thus, in modern <u>Xenopus laevis</u>, an ancestrally tetraploid species, mutants at many diploid loci have been discovered by inbreeding (Gurdon and Woodland, 1975) and by gynogenesis (Tompkins and Krotoski-Gwozdziowski, 1983). The purpose of this paper is to examine a few of the effects of two rather simple genetically based metrical changes affecting cell size and/or cell numbers in <u>Xenopus laevis</u>, tetraploidy and the allometric mutant <u>enlarged eye</u>, on the structure and function of the visual system.

SOME EFFECTS OF POLYPLOIDY ON NEURAL DEVELOPMENT

Polyploid amphibia develop normally by external criteria and do not appear to be defective in their general vigour. However, Fankhauser, et al. (1955) have shown that polyploid salamanders learn more slowly than do diploid animals, perhaps because their brains, although of normal size, contain fewer cells and the possible interactions between neurons is thus reduced. The metrical relationships between the Mauthner's neurons (identifiable giant neurons of the medulla of tailed lower vertebrates) and the rest of the nervous system is perturbed in polyploids because in diploid and polyploid salamanders the number of Mauthner's neurons remains two despite the reduction in cell number in the nervous system generally. It is also conceivable that polyploidy causes abnormalities in cell morphologies and function. The organ sizes of polyploid animals are similar to those of diploids and some neural cells have functions which require the same spatial arrangements relative to the organs in both diploid and polyploid animals (for example, the tectal radial glial cells span the tectum from the ventricle to the tectal surface). Therefore, increased cell size in polyploid animals may induce abnormalities in cell structure and function more complex than simple magnification of cell size and reduction in cell numbers would imply. We have investigated this using our stable autotetraploid line of <u>Xenopus laevis</u> and normal diploid animals. In addition, we have investigated the effects of interactions between diploid and tetraploid neural cells in chimerae constructed of tissues of both types during embryonic development.

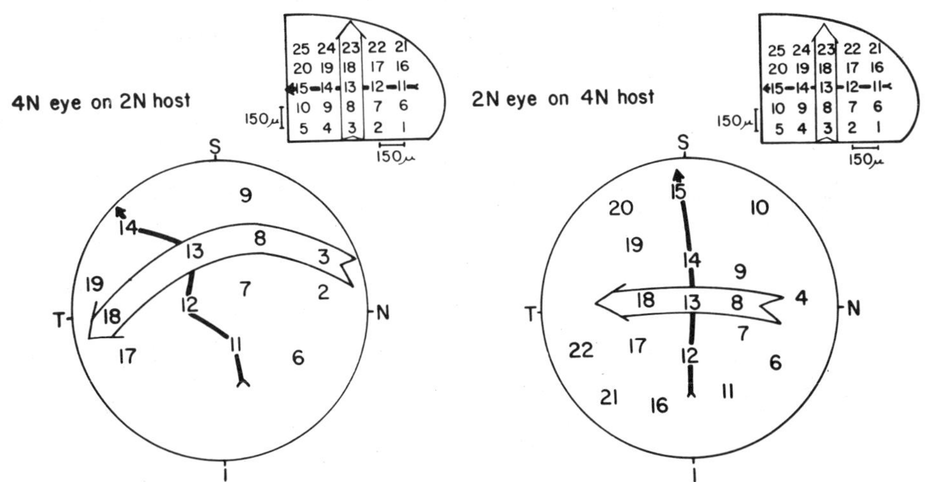

Fig. 1. Visuotectal projections of the right eyes of two diploid/ tetraploid chimeric frogs. Right eyes of diploid and tetraploid embryos were exchanged surgically at stage 32, just after the formation of the eye cup. The chimerae were reared through metamorphosis and then subjected to visuotectal mapping. The numbered positions in the illustrated contralateral tecta were tested with a platinum-iridium electrode and the position(s) in the visual field which when illuminated with a spot of light caused a recordable stimulation of the electrode were marked with the tectal position. Each circle represents 180° of visual field. Both the tetraploid eye on the diploid host and the diploid eye on the tetraploid host gave visuotectal projections which are essentially normal showing that diploid and tetraploid eyes and tecta can integrate functionally.

The same cell types found in diploid eyes and brains are found in the eyes and brains of tetraploid animals. Sakaguchi, et al. (in press) have shown that the same classes of retinal ganglion cells are found at the same developmental stages in diploid and tetraploid embryos and tadpoles. Visuotectal mapping has shown that the vision of tetraploid frogs is essentially the same as that of diploid frogs. The visuotectal projections of diploid eyes on tetraploid host animals and tetraploid eyes on diploid hosts are normal (Figure 1). Thus, at this level of assay the function of the tetraploid nervous system is normal and diploid and tetraploid eyes and brains can be functionally integrated.

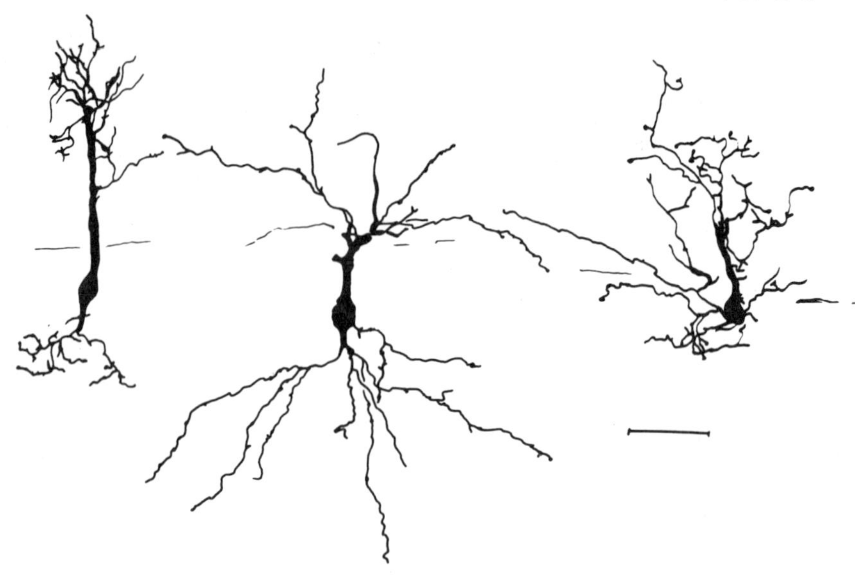

Fig. 2.   Camera lucida drawings of three large pear-shaped neurons
           from the tecta of a diploid animal (left), a tetraploid
           animal (center), and a tetraploid animal whose tectum was
           innervated by a diploid eye (right). These preparations,
           using the rapid golgi method modified from Somogyi, et al.
           (1979), were made after metamorphosis. The tetraploid cell's
           dendrites are much longer than those of the diploid cell.
           In both diploid and tetraploid cells in intact animals the
           basal dendrites project ventrally. In tetraploid cells of
           tecta innervated by diploid eyes these basal dendrites
           curve dorsalward. Despite this unusual morphology, diploid
           eyes innervate tetraploid tecta to give normal visuotectal
           projections. Bar = 50 microns.

    Analysis of golgi stained tectal cells of diploid and tetraploid
frogs shows that the increased cell size in the latter is expressed
in different ways by different cell types. Preservation of the diploid nucleocytoplasmic ratio and simple doubling of the cell sizes
in tetraploids would lead to an increase in linear dimensions of
1.26 times the linear dimensions of diploid cells. Tetraploid
neural cells are larger than their diploid counterparts, but the relationships between their parts are altered in qualitative ways
which are not predicted by simple magnification. For instance, the
large pear-shaped neurons in tetraploid tecta typically have dendrites which are considerably longer than 1.26 times their diploid

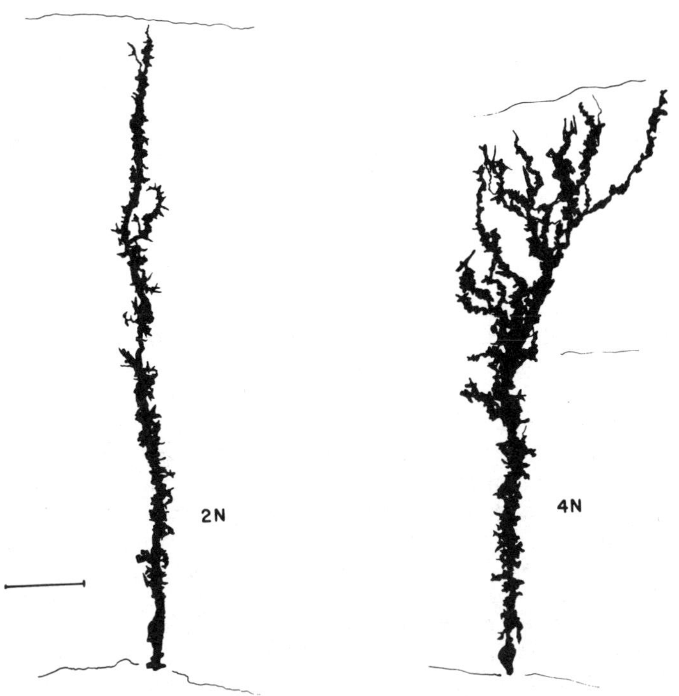

Fig. 3. Camera lucida drawings of golgi stained radial glial cells
of the tecta of postmetamorphic diploid (left) and tetraploid (right) animals. These glial cells span the tectum
from the ventricle to the surface of the tectum, a distance
not significantly different in diploid and tetraploid animals. The tetraploid cells are much more branched than are
the corresponding diploid cells. As in the case of the
pear-shaped neurons, tetraploidy does not merely magnify
the size of each type of cell. Rather, qualitative alterations in the relationships of the various parts of the
cells are evident. Bar = 50 microns.

counterparts (Figure 2). In contrast, the tectal radial glial cells
of tetraploid animals are probably constrained by the thickness of
the tectum, since they span the thickness of similar sized organs
in both diploids and tetraploids. Tetraploid radial glial cells
typically show more extensive branching than do diploid radial glial
cells (Figure 3). Thus, different tetraploid cell types express
their increased cell size in different ways. Tetraploid cells are
not merely magnified diploid cells but rather they express qualitative alterations in their morphologies. Golgi stained preparations
of tetraploid tecta innervated by diploid eyes show a further qualitative alteration in cell morphology (Figure 2). The large pear-
shaped neurons normally have basal dendrites which course ventrally

in both diploid and tetraploid animals. When innervated by diploid
eyes, tetraploid cells develop many basal dendrites which project
dorsalward. The reasons for this and the functional effects, if any,
are unknown. It does suggest that allometric alterations in cell
size and number of one member of a pair of interacting neural cell
types can alter the interactions and dendritic morphologies of these
cell types.

Ocular dominance columns can be induced in frog tecta by duplicate innervation of a tectum by an additional eye (Levine and
Jacobson, 1975), by similar half eyes in a compound eye such as a
double nasal or double temporal compound eye (Fawcett and Willshaw,
1982), or by a retinal fragment which regenerated a duplicated visuotectal projection (Ide et al., 1983). The morphology of the stripes
can be revealed by ablating that part of the retina giving rise to
one of the duplicated projections and labelling the remaining one
with 3H-proline. The stripes result from alternating innervation of
adjacent tectal positions by corresponding parts of each of the duplicated projections. They appear to be maintained by neural activity
(Meyer, 1982) and their width may reflect the width of the terminal
arborizations of the retinal ganglion cell fibers. Figure 4 shows
camera lucida drawings of sections of four tecta with duplicate innervation, one projection of each pair labelled with 3H-proline.
Figure 4A shows typical stripes in a diploid animal. In most preparations of diploid _Xenopus_ the number of dark stripes, corresponding
to the 3H-proline, is between six and eight. Figure 4B shows typical
tectal stripes in tetraploid animals. The number of stripes varies
from one to four and these are typically broader than those seen in
diploid animals. Figure 4C shows the labelled diploid projection in
a diploid nasal right/tetraploid nasal left compound eye innervating
a diploid tectum. Figure 4D shows the same type of animal with the
tetraploid projection labelled. In both the latter figures, the dis-
parity in stripe width is apparently preserved in competition with
fibers of the other ploidy. Thus, tetraploid retinal fibers form
stripes which are broader and fewer in number than do diploid retinal
fibers. This may reflect an increased terminal arbor size in tetraploid retinal ganglion cells. These results show that the physiological integration of tetraploid neurons leads to different patterns
of innervation. This change under abnormal experimental circumstances
may be reflected in other alterations in the patterns of physiological integration in other places in tetraploid brains.

SOME EFFECTS OF ALLOMETRIC CHANGES IN GROWTH OF NEURAL STRUCTURES

The recessive mutant _enlarged eye_ develops normally until the
onset of metamorphosis, after which mutant eyes grow more rapidly
than normal. By three months after metamorphosis the mutant eyes
are nearly twice the size of normal eyes on similar sized normal animals. Such mutant eyes contain about 1.4 times the normal number

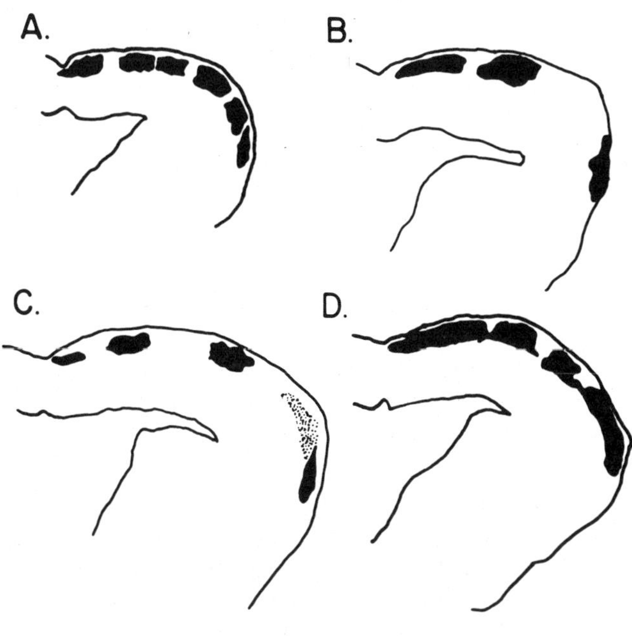

Fig. 4. Camera lucida drawings of autoradiographs of dominance
stripes in the tecta of animals whose contralateral eyes
were surgically perturbed to generate the stripes by virtue
of their duplicated retinotectal projections. After visuotectal mapping, the fibers of one projection were ablated
and the eye injected with 3H-proline. After 24 hours, the
animals were sacrificed and processed for autoradiography.
Figure 4A shows the stripes in a diploid animal's left tectum; innervation was by duplicate diploid projections.
Figure 4B shows the stripes in a tetraploid animal's left
tectum innervated by duplicate tetraploid projections.
Tetraploid stripes are fewer in number and broader than diploid stripes. Figures 4C and 4D show sections of the left
tecta of diploid animals whose right eye consisted of diploid nasal right and tetraploid nasal left halves. In C
the diploid cells of the retina were labelled with 3H-pro-
line; in D the tetraploid retinal cells were labelled. The
diploid and tetraploid stripes maintain their relative
widths in competition with one another. Bar = 200 microns.

of retinal ganglion cells, yet the visuotectal projections of mutant
eyes _in_ _situ_ are normal. The excessive growth of mutant eyes is expressed autonomously when transplanted to normal hosts and such eyes
have normal visuotectal projections.

The growth of normal eyes is not symmetrical during development.
The dorsal eye grows most during early development; near metamorphosis the ventral eye grows most (Hollyfield, 1971; rev., Jacobson,
1978). Despite this complex growth pattern which has evolved, normal
visuotectal projections can be recorded during tadpole and subsequent
stages of development. The tectal targets of the retinal projection
grows roughly in concert with the eye growth pattern during early
development. The lateral tectum is born during early development but
the mediocaudal tectum continues to grow through metamorphosis. The
lateral tectum receives the dorsal retinal fibers and the mediocaudal
tectum receives the ventral and nasoventral retinal fibers. In addition to the rough coincidence of the growth of the retina and its
tectal target, sliding connections on the tectum probably also play
a role in shaping the visuotectal map (Jacobson, 1976; Scott and Lazar, 1976; Gaze et al., 1979; Fraser, 1983).

Although _ee/ee_ eyes give normal visuotectal projections when intact, chimeric eyes composed of both mutant and normal retinal tissue
have unusual alterations in growth patterns analogous to those established in normal eyes during evolution. Mutant/normal double nasal
chimeric eyes have radically different visuotectal projections from
diploid/diploid or diploid/tetraploid chimerae of the same type.
For example, the duplicated visuotectal projection of a diploid nasal
right/tetraploid nasal left eye on a diploid host is shown in figure
5 (left). Figure 5 (right) shows the visuotectal projection of a
mutant nasal left/normal diploid nasal right chimeric eye on a normal
host. Instead of the usual duplicated projection, some of the latter
projection is triplicated. The abnormal triplication occurs in the
dorsal retina of the mutant genotype and is seen in the inferior
temporal visual field. Chimeric eyes resulting from heterotopic
grafts performed at early eyecup stages of development sometimes
result in visuotectal projections which do not reflect the positions
of origin of the grafted cells (Gaze, et al., 1983). Conway et al.
(1980) have proposed that this phenomenon results from alterations
in positional information specifying relative positional affinities
of the fibers on the tectum under the influence of and in conformance
to the grafted tissue's new surround. Cooke and Gaze (1983) have
proposed that intercalation of positional values may play a role in
certain "regulated" chimeric eyes. However, the triplicated projection produced by mutant/normal double nasal chimeric eyes reflects
an alteration in the dorsonasal retina of mutant origin in a dorso-
temporal position in the orbit leading those retinal cells to project to the tectal positions normally innervated by the nasoventral
retina. This alteration bears little relationship to the sort of
regulation to conform to the surround discussed by Conway et al.
(1980) or to intercalation. The key point may be that the mediocaudal tectum is growing and may provide open innervation sites for
the abnormally growing nasodorsal mutant retina. Since whole mutant

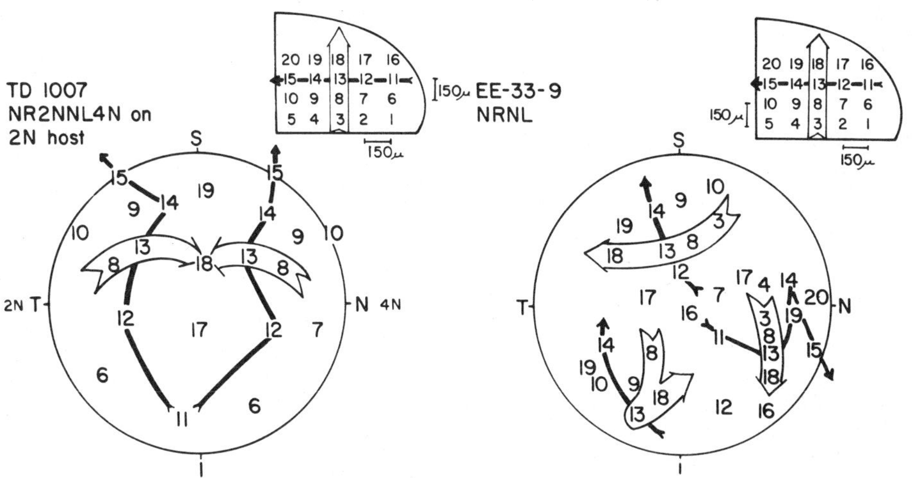

Fig. 5.   Visuotectal maps of compound nasal left/nasal right eyes in
          the right orbits of normal diploid hosts. On the left, the
          nasal left portion was tetraploid and the nasal right was
          normal diploid. In the animal whose visuotectal projection
          is shown on the right the nasal left was homozygous for the
          enlarged eye mutant and the nasal right was normal diploid.
          The left map is a simple duplicate. Duplicate points for
          central positions 11, 17, 18, and 19 could not be discrimin-
          ated by the assay. The map on the right shows a triplicated
          sector in the inferior temporal visual field which is not
          seen in compound double nasal projections of diploid/diploid
          or diploid/tetraploid chimeric eyes. This triplication
          corresponds to retina of the mutant genotype in the dorsal
          portion of the nasal left retinal half. The points which
          are triplicated are in the mediocaudal sector of the tectum,
          close to the tectum's growing edge.

eyes have normal visuotectal projections, it may be that the abnormal
growth relationships in the dorsal retina resulted in the abnormal
projection of the dorsal fibers to the mediocaudal tectum. The induction of unusual fiber projections resulting from local changes in
growth patterns in this abnormal experimental circumstance probably
reflects the importance of coordinate growth patterns of interacting
organs during development to the formation of proper connections.
Allometric growth therefore may provide an evolutionary mechanism
for the development of new pathways within the brain. Such a mechanism does not initially appeal to alterations in fine genetic control
of development, but if the new pathways are advantageous, they may
be brought under such control in subsequent evolution.

CONCLUSIONS

The genetic bases for the alterations in cell size and cell
number caused by autotetraploidy and by the mutant _enlarged eye_ are
relatively simple in nature. Tetraploidy and the _enlarged eye_ mutation by themselves result in frogs which are viable and superficially
normal. However, when examined in finer detail, or when new interactions are created by the construction of chimeric animals, fundamental and startling effects are revealed.

Tetraploidy alone alters neural dendritic morphology. Tetraploid
neurons generally have either longer or more dendritic branches than
their diploid counterparts. These effects could underlie alterations
in neurophysiological interactions between neurons which might ultimately affect behavior or learning ability. In chimeric animals,
tetraploid neurons are brought into new interactions with either
other tetraploid cells or with diploid cells. Innervating a tetraploid tectum with a diploid eye results in new dendritic configura-
tions. Tectal striping, which has been touted as a model system for
mammalian ocular dominance columns, reveals new patterns of axon
arbor domains.

The _enlarged eye_/normal chimeric eyes produce an even more interesting phenomenon. The mutant dorsal retina produces an entirely
new projection to the growing mediocaudal edge of the tectum. This
raises the possibility that alterations in growth rates of either the
target organ or the projecting neurons could form the basis for new
projections in the brain. Admittedly, these competitive interactions
have been constructed artificially. However, if changes in cell
size, number, or growth rates were to occur in a single cell type or
restricted domain of the brain, new projections or cell morphologies
could conceivably occur without the intervention of the embryologist.

These relatively simple allometric alterations suggest that
allometric changes in general might lead to the development of new
pathways, and concomitantly, perhaps new functions. Thus, simple
metrical alterations such as those illustrated by tetraploidy and
the _enlarged eye_ mutation may have played important roles in the
evolution of the structure and function of the nervous system.

AKNOWLEDGEMENTS

This research was supported by National Science Foundation
grants PCM-82-09293 and BNS-82-16681. B. Szaro was supported by a
National Research Service Award fellowship 5 F 32 NS06948 from the
National Institute of Neurology and Communicative Disorders and
Stroke.

REFERENCES

Bogart, J. P., (1979). Evolutionary implications of polyploidy in
    amphibians and reptiles, in: "Polyploidy: Biological Relevance",
    W. H. Lewis, ed., Plenum Press, New York, 341-378.
Conway, K., Feiock, K., and Hunt, R. K., (1980). Polyclones and
    patterns in growing Xenopus eye. Curr. Top. Develop. Biol.
    15:217-317.
Cooke, J., and Gaze, R. M., (1983). The positional coding system in
    the early eye rudiments of Xenopus laevis, and its modification
    after grafting operations. J. Embryol. Exp. Morph. 77:53-71.
Fankhauser, G., (1955). The role of nucleus and cytoplasm, in: "Anal-
    ysis of Development", B. H. Willier, P. A. Weiss, and V. Ham-
    burger, eds, W. B. Saunders, Philadelphia, 126-150.
Fankhauser, G., Vernon, J. A., Frank, W. H., and Slack, W. V., (1955).
    Effect of size and number of brain cells on learning in larvae
    of the salamander, Triturus viridescens. Science 122:692-693.
Fawcett, J. W., and Willshaw, D. J., (1982). Compound eyes project
    stripes on the optic tectum in Xenopus. Nature 296:350-352.
Fraser, S. E., (1983). Fiber optic mapping of the Xenopus visual sys-
    tem: shift in the retinotectal projection during development.
    Develop. Biol. 95:505-511.
Gaze, R. M., Keating, M. J., Ostberg, A., and Chung, S.-H., (1979).
    The relationship between retinal and tectal growth in larval
    Xenopus: implications for the development of the retino-tectal
    projection. J. Embryol. Exp. Morph. 53:103-143.
Gould, S. J., (1977). "Ontogeny and Phylogeny", Belknap Press, Cam-
    bridge.
Gurdon, J. B., and Woodland, H. R., (1975). Xenopus, in: "Handbook
    of Genetics", vol. 4, R. C. King, ed., Plenum Press, New York,
    35-50.
Hollyfield, J. G., (1971). Differential growth of the neural retina
    in Xenopus laevis larvae. Develop. Biol. 24:264-286.
Ide, C. F., Fraser, S. E., and Meyer, R. L., (1983). Eye dominance
    columns from an isogenic double nasal frog eye. Science 221:293-
    295.
Jacobson, M., (1976). Histogenesis of the retina in the clawed frog
    with implications for the pattern of development of retino-
    tectal connections. Brain Res. 103:541-545.
Jacobson, M., (1978). "Developmental Neurobiology", Plenum Press, New
    York.
Kobel, H. R., (1981). Evolutionary trends in Xenopus(Anura Pipidae).
    Monitore zoologico italiano 8:119-131.
Levine, R. and Jacobson, M., (1975). Discontinuous mapping of retina
    onto tectum innervated by both eyes. Brain Res. 98:172-176.
Meyer, R. L., (1982). Tetrodotoxin blocks the formation of ocular
    dominance columns and refined retinotopography in goldfish.
    Neurosci. Abstr. 8:436.
Ohno, S., (1970). "Evolution by Gene Duplication", Springer-Verlag,
    New York.

Reinschmidt, D. C., Simon, S. J., Volpe, E. P., and Tompkins, R.,
    (1979). Production of tetraploid and homozygous diploid amphib-
    ians by suppression of first cleavage. J. Exp. Zool. 210:137-143.
Sakaguchi, D. S., Murphey, R. K., Hunt, R. K., and Tompkins, R., in
    press. The development of retinal ganglion cells in a tetra-
    ploid strain of Xenopus laevis: a morphological study utilizing
    intracellular dye injection. J. Comp. Neurol.
Scott, T. M., and Lazar, G., (1976). An investigation into the hypoth-
    esis of shifting neuronal relationships during development,
    J. Anat. 121:485-496.
Somogyi, P., Hodgson, A. J., and Smith, A. D., (1979). An approach to
    tracing neuron networks in the cerebral cortex and basal gang-
    lia. Combination of golgi staining, retrograde transport of
    horse-radish peroxidase and anterograde degeneration of syn-
    aptic boutons in the same material. Neuroscience 4:1805-1852.
Tompkins, R., and Krotoski-Gwozdziowski, D., (1983). Mutants, cell
    markers, and inbred lines of Xenopus laevis for developmental
    studies. Genetics 104:s69.
Willshaw, D. J., Fawcett, J. W., and Gaze, R. M., (1983). The visuo-
    tectal projections made by Xenopus "pie slice" compound eyes.
    J. Embryol. Exp. Morph. 74:29-45.

THE MECHANISM OF BINDING OF

NEURAL CELL ADHESION MOLECULES

Stanley Hoffman and Gerald M. Edelman

The Rockefeller University
1230 York Avenue
New York, New York 10021

The morphology and cellular interconnections within the nervous
system of higher organisms arise as a result of complex cell-cell
interactions occurring throughout development. In order to identify the cell surface molecules mediating the adhesion that regulates
many of these interactions, a variety of assays have been used and
several molecules have been proposed to be involved in neural cell
adhesion (Merrell et al., 1975; Hausman and Moscona, 1976; Urushihara
et al., 1979; Marchase et al., 1981; Magnani et al., 1981; Grunwald
et al., 1982; Roth and Pierce, 1982; Sadoul et al., 1983; Lindner et
al., 1983). In our laboratory, an immunological approach devised
by one of us (GME) led to the identification (Brackenbury et al.,
1977; Thiery et al., 1977) and purification of a protein involved
in calcium-independent neural cell adhesion which has been named
the neural cell adhesion molecule or N-CAM. A number of other CAMs
of different structure and specificity have since been identified
(See Edelman, 1983, for a review).

N-CAM, isolated from embryonic chick brains, is a high molecular weight, sialic acid-rich, cell surface glycoprotein (Hoffman et
al., 1982). Following the enzymatic removal of the sialic acid,
two polypeptide chains of apparent molecular weight 140,000 and
170,000 can be resolved (Cunningham et al., 1983). Removal of all
N-linked carbohydrate reduces these molecular weights to 130,000
and 160,000. In a linear model of the N-CAM molecule, the bulk of
the sialic acid appears to be localized in a middle domain, between
the binding domain and the domain involved in anchoring the protein
to the cell surface (Cunningham et al., 1983). The sialic acid in
N-CAM appears primarily in the form of polysialic acid (Hoffman et
al., 1982; Finne et al., 1983). The two polypeptide chains in N-CAM

147

described above are closely related, having identical N-terminal
sequences and only minor differences in peptide fragments resolva-
ble following digestion with Staphylococcus aureus V8 protease
(Cunningham et al., 1983). The structure of N-CAM changes during
development. N-CAM, isolated from adult brains (Rothbard et al.,
1982; Edelman and Chuong, 1982) is similar to partially desialyla-
ted embryonic N-CAM on SDS gel electrophoresis. These analyses indicated that the polypeptide chains in adult N-CAM are very similar or identical to those in embryonic N-CAM, but that adult N-CAM
(A form) contains only 1/3 as much sialic acid as embryonic N-CAM
(E form).

The function and distribution of N-CAM have been studied primarily by using various antibody preparations directed against the
molecule to perturb its function. In addition to inhibiting neural
cell adhesion in vitro (Thiery et al., 1977), univalent Fab' fragments of rabbit anti-(N-CAM) antibodies inhibit the fasciculation of
neurite outgrowths from cultured spinal ganglia (Rutishauser et al.,
1978), the formation of histological layers in retinas developing
in organ culture (Buskirk et al., 1980), and adhesion between
neural and muscle cells in culture (Rutishauser et al., 1983). Using a crossreactive monoclonal antibody, N-CAM has been detected in
frog, chicken, mouse, rat, and human tissue (Chuong et al., 1982;
McClain and Edelman, 1982). Detailed immunofluorescence studies
on the developmental distribution of N-CAM have been carried out
in the chicken (Thiery et al., 1982; Edelman et al., 1983; Thiery
et al., in press).

N-CAM is present on all neurons in both the central and peripheral nervous systems from their formation and throughout the life
of the animal and is found on striated and heart muscle throughout
embryogenesis. The distribution of N-CAM during the earliest stages
of chicken embryogenesis is particularly provocative. N-CAM can
be detected throughout the entire embryo at the blastoderm stage.
Furthermore, N-CAM is present transiently at several sites where
inductive events occur suggesting a link between the fluctuation in
N-CAM concentration and embryonic induction (Edelman et al., 1983).

Monoclonal antibodies that distinguish between the E and A forms
of N-CAM have been used to analyze E to A conversion in perinatal
mouse brain (Edelman and Chuong, 1982). It was found that the E
to A conversion in N-CAM structure occurs at different rates and
to different extents in various normal brain regions. Of particular significance is the fact that E to A conversion is delayed in
the cerebellum of the neurological mutant, staggerer (Edelman and
Chuong, 1982).

There are two basic hypotheses concerning the nature of the
cell surface proteins that mediate pattern formation in the developing nervous system. The strict chemoaffinity hypothesis (Sperry,

148

1963) suggests that specificity results from the presence of a large
number of narrowly-localized, highly-specific, complementary markers on interacting cells. On the other hand, the modulation hypothesis (Edelman, 1983) proposes that a relatively small number
of adhesive proteins (of the order of dozens) could mediate pattern
formation provided that their functions can be more or less continuously graded by such factors as chemical modification or variations in temporal or spatial expression. In the remainder of this
paper a series of N-CAM binding experiments are described which
support the modulation hypothesis by demonstrating that the binding efficacy of N-CAM on membranes is strongly affected by two processes known to occur in vivo: differential sialylation (i.e. E
to A conversion) and changes in local surface concentration. The
major conclusions of these studies are that i) N-CAM-mediated adhesion is homophilic, i.e. N-CAM to N-CAM; ii) the rate of N-CAM-
mediated adhesion is inversely related to the sialic acid content
of the N-CAM molecules; and iii) the rate of N-CAM-mediated adhesion is highly dependent on the local surface concentration of
N-CAM.

MOLECULAR MECHANISM OF N-CAM BINDING

     N-CAM was first identified as being involved in cell adhesion
on the basis of its ability to neutralize the neural cell adhesion
inhibiting activity of polyspecific antisera directed against neural
antigens (Thiery et al., 1977). This identification, however, gave
no clue as to the molecular mechanism of neural cell adhesion and
did not prove that N-CAM is itself a direct ligand in cell-cell
binding. The results reviewed here strongly indicate that N-CAM
is a ligand in neural cell adhesion and that its receptor on the
adhering cell is a second N-CAM molecule. Its binding mechanism
is therefore second-order homophilic (see Edelman, 1983).

     Several experiments have suggested a specific binding of N-CAM
to cells. As shown in Table 1, soluble radioiodinated N-CAM was
found to bind to retinal cells and myoblasts. The specificity in-
herent in this binding was demonstrated by the fact that unlabeled
soluble N-CAM and anti-(N-CAM) Fab' fragments inhibited this binding and that little, if any, binding to fibroblasts, hepatocytes,
or trypsinized retinal cells was detected. A more graphic demonstration of N-CAM binding was obtained using fluorescently-labeled
reconstituted lipid vesicles containing N-CAM at their surface.
These vesicles bound well to retinal cells (Fig. 1) and muscle cells
(Grumet et al., 1982) and this binding was strongly inhibited by
anti-(N-CAM) Fab' fragments. As a further specificity control,
vesicles lacking N-CAM did not bind to cells (data not shown).

     Several correlations observed in these binding experiments
suggested that the cellular receptor for soluble N-CAM or N-CAM in

Table 1. Binding of soluble N-CAM to cells

Cell Type	Inhibitor or competitor*	Bound N-CAM, cpm[+]		
Retina	——	23,700	±	100
Retina	Anti-(N-CAM)	700	±	300
Retina	N-CAM	7,200	±	1,000
Retina (trypsinized)	——	2,500	±	500
Myoblasts	——	11,200	±	1,100
Myoblasts	Anti-(N-CAM)	2,500	±	500
Fibroblasts	——	3,200	±	200
Hepatocytes	——	2,300	±	300

*Anti-(N-CAM) was 20 µg of affinity-purified or monoclonal anti-(N-CAM) Fab'. N-CAM was 100 µg of the unlabeled molecule. [+]After incubation of 1×10^7 cells for 15 min. at 25° C. in 200 µl of medium containing 3.5 mg bovine serum albumin and 2 µg of ^{125}I-N-CAM (2×10^5 cpm). Values are the mean ±SD of three experiments.
Data from Rutishauser et al., 1982.

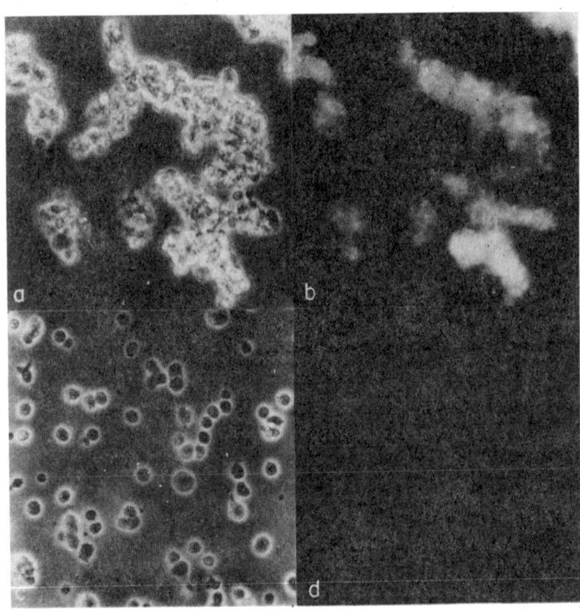

Fig. 1. Binding of fluorescent N-CAM vesicles to retinal cells. (a) Phase-contrast micrograph. (b) Fluorescence micrograph of the same field. (c and d) Same experiment carried out in the presence of anti-(N-CAM) Fab'. (from Rutishauser et al., 1982).

vesicles was N-CAM itself. Binding was observed to cells bearing
surface N-CAM (retinal cells, brain cells, and myoblasts) but not
to cells lacking N-CAM (hepatocytes, fibroblasts, and freshly
trypsinized retinal cells). Furthermore, it was observed that
N-CAM containing vesicles self-aggregate and that this aggregation
is blocked by anti-(N-CAM) Fab' fragments (Fig. 2). Vesicles made
without N-CAM did not aggregate. Taken together, these results
can be explained by a molecular mechanism for aggregation in which
N-CAM binds to N-CAM.

To test this hypothesis, additional binding experiments were
performed in which retinal cells were reacted with anti-(N-CAM) or
nonimmune Fab' fragments and washed before use. If the binding of
exogenous N-CAM were to a cellular receptor other than N-CAM, pre-
treatment of the cells with anti-(N-CAM) Fab' fragments should have
no effect in N-CAM binding experiments because the cellular receptor would not be blocked. This, however, was not the case. Pre-
treatment of retinal cells with anti-(N-CAM) Fab' fragments strongly inhibited binding of both soluble N-CAM and N-CAM vesicles (Table
2).

MODULATION OF N-CAM BINDING

The experiments described above were useful in demonstrating
that N-CAM binds to cells and in suggesting that the molecular
mechanism of this binding is N-CAM to N-CAM. Experiments of this
kind were not particularly suited, however, for establishing the
quantitative mechanisms of binding nor for evaluating the functional
significance of the dynamic changes in N-CAM structure and expression that have been observed in vivo. During the perinatal period
in mice, for example, the sialic acid concentration in N-CAM molecules gradually decreases down to a level only 1/3 that found in
embryonic N-CAM (Edelman and Chuong, 1982). The amount of N-CAM
expressed in certain early embryonic tissues is also dynamic (Thiery
et al., 1982; Edelman et al., 1983). In perhaps the most interesting case, neural crest cells express N-CAM when they first appear
at the dorsal border of the neural tube, lose most or all of their
N-CAM concomitant with migration, and finally, upon reaching their
destination, re-express the molecule while clustering and differentiating into ganglionic precursors. Thus there is a correlation
in vivo between the presence of N-CAM on cells and their state of
aggregation. These and other results have led to the proposal that
pattern formation may result from modulation of the function of
N-CAM and other cell adhesion molecules by such factors as chemical
modification and differential gene expression (Edelman, 1983;
Edelman, in press).

To directly test the modulation hypothesis in the case of N-CAM,
it was necessary to design a quantitative kinetic adhesion assay in

Table 2. Binding properties of retinal cells coated with Fab'

Fab' used to coat cells*	N-CAM + vesicles	Soluble N-CAM, cpm[†]
Nonimmune	+++	30,100 ± 3,100
Anti-(N-CAM)	+/-	2,000 ± 200

*Monoclonal or affinity-purified Fab' was used at 20 μg/ml,
 and unbound Fab' was removed by washing.
⧾Visual estimation of bound fluorescent vesicles.
†See Table 1 for details.
 Data from Rutishauser et al., 1982.

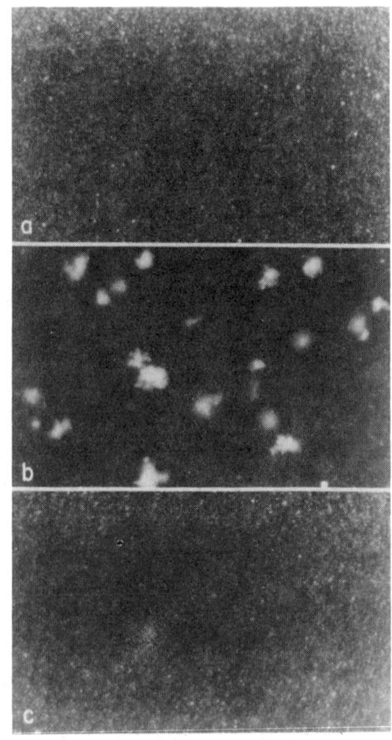

Fig. 2. Aggregation of N-CAM vesicles as observed by fluorescence
        microscopy. (a) Freshly prepared vesicles. (b) Vesicles in-
        cubated with shaking for 2 hr at 37°C. (c) Vesicles incuba-
        ted as in (b) in the presence of affinity-purified
        anti-(N-CAM) Fab' at 20 μg/ml. (from Rutishauser et al.,
        1982).

which the structure and concentration of N-CAM could be experimentally manipulated. Cells are not appropriate for these strictly quantitative studies because i) they cannot withstand prolonged enzymatic treatments at non-physiological pH and ionic strength, ii) during many cell preparation protocols the surface density and structure of N-CAM found in the parent tissue are altered, and iii) an exact estimate of surface concentration of CAMs is difficult. On the other hand, as described above, reconstituted lipid vesicles containing N-CAM retain the adhesive specificity of the cells from which the molecules were prepared, will self-aggregate, can be prepared containing defined N-CAM concentrations, and are still functional following enzymatic treatments. Therefore, in the kinetic adhesion assay the aggregation of reconstituted lipid vesicles containing N-CAM was studied. Parameters analyzed for their relationship to rate of aggregation included N-CAM to lipid ratio in the vesicles, sialic acid concentration on the N-CAM molecules, and a comparison of the E and A forms of N-CAM. To confirm the results obtained with reconstituted vesicles in a more native system, similar experiments were performed using membrane vesicles prepared directly from brain tissue homogenates; these vesicles could be used as a control for the effects of other cell surface proteins on the kinetic assay.

In the kinetic adhesion assay, the growth of vesicle aggregates was monitored using a Coulter Counter set to count particles larger than 1.5 μm^3, its lowest reliable threshold. Rates of aggregation were calculated from the increase with time of total volume of vesicles in superthreshold aggregates. Rate constants indicating the relative adhesiveness of a vesicle preparation could then be calculated from the equation $V=K_{agg}A^2$ where Kagg is the apparent rate constant, V is the rate of aggregation measured at vesicle concentration A, and the exponent is the experimentally derived dependence of rate of aggregation on initial vesicle concentration.

When this assay was used to compare the aggregation of reconstituted vesicles containing different N-CAM concentrations, it was found that a two-fold increase in N-CAM concentration resulted in a greater than 30-fold increase in apparent rate constant for aggregation (Table 3) with intermediate rates observed at intermediate N-CAM concentrations. This result is equivalent to a fifth-order dependence of adhesivity on local N-CAM concentration in a membrane and raises the possibility that small variations in spatial or temporal expression of N-CAM _in vivo_ could have major effects on pattern formation. The high-order dependence of rate of aggregation on _N-CAM concentration in_ vesicles should not be confused with the second-order dependence of rate of aggregation on _vesicle concentration_ mentioned above. It is also important to note that these results indicate that extreme care must be taken to normalize surface concentration before considering the effect of other parameters such as chemical modification on adhesive function.

Table 3. Effect of N-CAM-to-lipid ratio on vesicle aggregation

N-CAM/lipid, µg/mg	Kagg, units*
14.0	1.5
16.9	3.5
18.8	6.4
28.3	54.0

*One unit is the rate of aggregation of a sample (measured in nl
of superthreshold product per ml per min) divided by the square
of the concentration of vesicles in the sample (vesicle concentration was measured in mg of lipid per ml).
Data from Hoffman and Edelman, 1983.

   In order to compare the function of the structurally distinct
forms of N-CAM purifiable from embryonic and adult brain tissue,
reconstituted vesicles containing very similar concentrations of
the E and A forms of N-CAM were prepared and their aggregation
analyzed. Vesicles containing the A form of N-CAM aggregated with
an apparent rate constant 3.5 times that of vesicles containing
the E form of N-CAM. Previous studies have suggested that the
primary structural difference between the E and A forms of N-CAM
is in their sialic acid content (Rothbard et al., 1982). To evaluate whether this difference in sialic acid content is also the
source of the functional difference, embryonic or adult N-CAM vesicles containing a range of sialic acid concentrations on their

Table 4. Effect of neuraminidase treatment on vesicle aggregation

Vesicles	Neuraminidase treatment	Sialic acid removed, %	Relative Kagg
E form	Buffer only	<5	1.0
	0.025 unit/ml	50	2.3
	0.1 unit/ml	70	3.1
	0.4 unit/ml	>95	3.7
A form	Buffer only	<5	1.0
	0.4 unit/ml	>95	1.1

E form and A form N-CAM vesicles at a concentration of 4.2 mg/ml
of lipid were neuraminidase-treated for 1 hour at 37°C using the
indicated concentrations of enzyme. Vesicles were then washed and
the release of sialic acid and rate of aggregation quantitated for
each preparation. Kagg values were normalized to 1.0 for buffer-
treated vesicles.
Data from Hoffman and Edelman, 1983.

N-CAM molecules were produced by differential neuraminidase treatment (Table 4). Complete removal of sialic acid from E form N-CAM vesicles enhanced their rate of aggregation 3.7-fold with intermediate rates observed at intermediate sialic acid concentrations. Almost all of this enhancement could be obtained by 70% removal of sialic acid to a level similar to that in A form N-CAM. On the other hand, complete removal of sialic acid from A form N-CAM vesicles only marginally enhanced their rate of aggregation. These results indicate that there is an inverse relationship between the sialic acid content and the adhesive function of N-CAM but that nearly maximal rates of aggregation are already attained at the sialic acid concentration found in typical A forms.

Although the above results indicate that vesicles containing A form N-CAM self-aggregate more rapidly than E form N-CAM vesicles, they do not address the question of whether A and E forms of N-CAM will interact with each other and, if so, with what relative strength. To answer this question, the aggregation of mixtures of A form N-CAM vesicles and E form N-CAM vesicles was monitored (Fig. 3). The fact that the total rate of aggregation in each mixture was greater than the sum of the expected contributions of E form N-CAM vesicle to E form N-CAM vesicle aggregation and A form N-CAM vesicle to A form N-CAM vesicle aggregation indicates that E form N-CAM vesicle to A form N-CAM vesicle aggregation must be occurring. Using an expanded form of the rate equation (Hoffman and Edelman, 1983), it can be calculated that the effective rate constant for E form N-CAM vesicle to A form N-CAM vesicle adhesion is intermediate between those for E form to E form N-CAM vesicle adhesion and A form to A form N-CAM vesicle adhesion.

The dependencies of N-CAM adhesive function on the sialic acid concentration in N-CAM and local N-CAM prevalence were confirmed in vesicle aggregation experiments performed using native membrane vesicles prepared directly from homogenates of embryonic and adult brain tissue. These vesicles also aggregate primarily by an (N-CAM)-mediated mechanism as demonstrated by a greater than 90% inhibition of their rate of aggregation by Fab' fragments of antibodies to N-CAM. As with the reconstituted vesicles, neuraminidase treatment greatly enhances the aggregation of embryonic native vesicles, but has only a marginal effect on adult native vesicles (Table 5). The effect of N-CAM concentration on native vesicle aggregation can be analyzed in the absence of differences due to sialic acid content by comparing the aggregation of neuraminidase-treated embryonic and neuraminidase-treated adult native vesicles. In this case (Table 5), the embryonic vesicles (which contain about 50% more N-CAM per mg of membrane protein) aggregate 3.6 times as rapidly as adult vesicles. Therefore, as in the case of reconstituted vesicles, there is a high-order dependence of rate of aggregation on local N-CAM concentration.

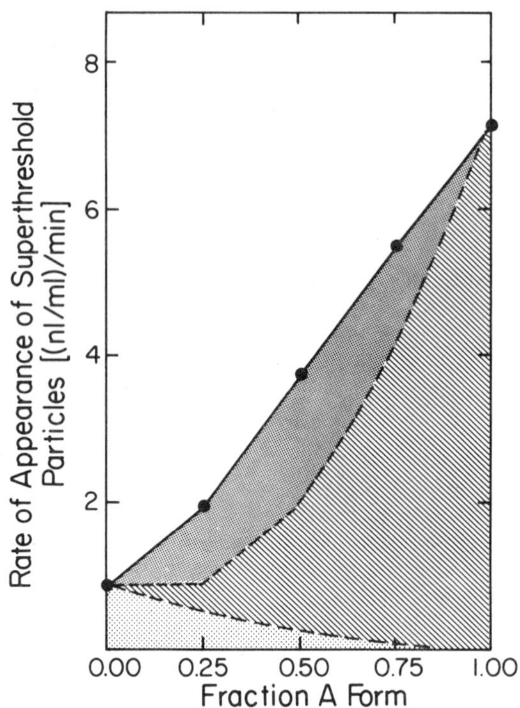

Fig. 3 Mixing experiment. E-form vesicles (14.0 µg of N-CAM per
mg of lipid) and A-form vesicles (16.3 µg of N-CAM per mg
of lipid) were coincubated at the five indicated propor-
tions with a constant total of 0.8 mg of vesicle lipid per
ml in each mixture. The apparent initial rate of aggregation of each mixture is plotted (●) versus the fraction of
A form. The calculated contributions of E-E (lightly
stippled) and A-A (hatched) interactions in each mixture
were obtained by multiplying the rate of aggregation of a
pure population of E-form vesicles or A-form vesicles by
the square of the relative concentration of each vesicle
species in a mixture in order to reflect the second-order
dependence of the rates on vesicle input. For example, for
the fraction A form = 0.25, the A-A contribution would be
$(0.25)^2$ x the apparent rate of aggregation of the pure
A-form population, and the E-E contribution would be $(0.75)^2$
x the rate of aggregation of the pure E-form population.
The residual aggregation in each mixture (heavily stippled)
is attributable to E-A interaction. (from Hoffman and
Edelman, 1983).

Table 5. Aggregation of native vesicles

Incubation	Kagg, units*	
	E Form	A form
Buffer only	1,810	2,200
Neuraminidase	9,230	2,560

E form and A form native vesicles at a concentration of
4 mg/ml of membrane protein were incubated for 3 hours at
37°C. in neuraminidase treatment buffer alone or containing 1 unit of neuraminidase per ml. Vesicles were then
washed and their aggregation analyzed.
*For native vesicle experiments, one unit is the rate of
aggregation of a sample (measured in nl of superthreshold
product per ml per min) divided by the square of the
concentration of vesicles in the sample (vesicle concentration was measured in mg of membrane protein per ml).
Data from Hoffman and Edelman, 1983.

SUMMARY

The experimental results reviewed in this paper strongly suggest
that the molecular mechanism of N-CAM-mediated cell adhesion involves the direct interaction of N-CAM molecules on one cell with
N-CAM molecules on a second cell. The rate of this aggregation has
a high-order dependence on the local N-CAM concentration, and is
inversely related to the sialic acid content of the N-CAM molecules
involved. In accordance with their relative sialic acid concentrations, the relative rates of aggregation mediated by E and A forms
of N-CAM are A-A > A-E > E-E. Further removal of sialic acid from
N-CAM below the level found in the A form gives little further enhancement of aggregation.

These results provide one basis upon which to interpret the
modulation hypothesis (Edelman, 1983) for control of N-CAM function,
i.e. the adhesive strength of N-CAM bonds in an in vitro system can
be altered in a graded manner over a wide range by variations in the
local surface density of N-CAM or by chemical modification of N-CAM
(differential sialylation). It is important to stress that these
results do not preclude the possibility of other forms of modulation
of N-CAM function or the function of other molecules in cell-cell
interactions.

It will be much more difficult to assess the role of N-CAM and
the modulation of its function on pattern formation in vivo. It
is pertinent to mention, however, that recent experiments on transformed neural cells (Greenberg et al., 1984) show loss of N-CAM

following transformation with accompanying loss of aggregation and
increased motility of the transformed cells. Aside from the possible implications for metastasis (transformation has for the first
time been shown to affect a defined CAM and alter cellular sociol-
ogy), these findings are consonant with the notion that alteration
of surface N-CAM affects expression of other cellular processes.
Clearly additional experiments are required to define the mechanisms by which this occurs. In addition to mapping the prevalence
and form of N-CAM during embryonic development, it will be necessary
to pruturb both its functions and the modulation of its function using
reagents that will discriminate among various forms of the molecule
in the embryo itself. These problems will require elegant solutions
but must be solved if completely satisfactory answers to questions
about the role of N-CAM _in vivo_ are to be obtained.

ACKNOWLEDGMENTS

This work was supported by U.S. Public Health Service Grants AI-11378,
HD-09635, HD-16550, and AM-04256 and by a fellowship to S.H. from
R.J. Reynolds Industries.

REFERENCES

Brackenbury, R., Thiery, J.-P., Rutishauser, U. and Edelman, G.M.
(1977) Adhesion among neural cells of the chick embryo. I. An
immunological assay for molecules involved in cell-cell binding.
J. Biol. Chem. 252:6835-6840.
Buskirk, D.R., Thiery, J.-P., Rutishauser, U., and Edelman, G.M.
(1980) Antibodies to a neural cell adhesion molecule disrupt
histogenesis in cultured check retinae. Nature 285:488-489.
Chuong, C-M., McClain, D.A., Streit, P., and Edelman, G.M. (1982)
Neural cell adhesion molecules in rodent brains isolated by
monoclonal antibodies with cross-species reactivity. Proc. Natl.
Acad. Sci. USA 79:4234-4238.
Cunningham, B.A., Hoffman, S., Rutishauser, U., Hemperly, J.J., and
Edelman, G.M. (1983) Molecular topography of N-CAM: surface
orientation and the location of sialic acid-rich and binding regions.
Proc. Natl. Acad. Sci. USA 80:3116-3120.
Edelman, G. M. (1983) Cell adhesion molecules. Science 219:450-457.
Edelman, G.M. and Chuong, C.-M. (1982) Embryonic to adult conver-
sion of neural cell adhesion molecules in normal and staggerer mice.
Proc. Natl. Acad. Sci. USA 79:7036-7040.
Edelman, G.M., Gallin, W. J., Delouvée, A., Cunningham, B.A., and
Thiery, J.-P. (1983) Early epochal maps of two different cell
adhesion molecules. Proc. Natl. Acad. Sci. USA 80:4383-4388.
Edelman. G.M. Cell adhesion and morphogenesis: The regulator
hypothesis. Proc. Natl. Acad. Sci. USA, in press.
Finne, J., Finne, U., Deagostini-Bazin, H., and Goridis, C. (1983)
Occurrence of α2-8 linked polysialosyl units in a neural cell
adhesion molecule. Biochem. Biophys. Res. Comm. 112:482-487.
Greenberg, M.E., Blackenbury, R., and Edelman, G.M. Alteration of
N-CAM expression following neuronal cell transformation by Rous
sarcoma virus. Proc. Natl. Acad. Sci. USA in press.

Grumet, M., Rutishauser, U., and Edelman, G.M. (1982) Neural cell adhesion molecule is on embryonic muscle cells and mediates adhesion to nerve cells in vitro. Nature 295:693-695.

Grunwald, G.B., Pratt, R.S., and Lilien, J. (1982) Enzyme dissection of embryonic cell adhesive mechanisms. III. Immunological identification of a component of the calcium-dependent adhesive system of embryonic chick neural retina cells. J.Cell Sci. 55:69-83.

Hausman, R.E., and Moscona, A.A. (1976) Isolation of a retina-specific cell-aggregating factor from membranes of embryonic neural retina tissue. Proc. Natl. Acad. Sci USA 73:3594-3598.

Hoffman, S. and Edelman, G.M. (1983) Kinetics of homophilic binding by E and A forms of the neural cell adhesion molecule. Proc. Natl. Acad. Sci. USA 80:5762-5766.

Hoffman, S., Sorkin, B.C., White, P.C., Brackenbury, R., Mailhammer, R., Rutishauser, U., Cunningham, B.A. and Edelman, G.M. (1982) Chemical characterization of a neural cell adhesion molecule purified from embryonic brain membranes. J. Biol. Chem. 257:7720-7729.

Lindner, J., Rathjen, F.G., and Schachner, M. (1983) L1 mono- and polyclonal antibodies modify cell migration in early postnatal mouse cerebellum. Nature 305:427-430.

Magnani, J.L., Thomas, W.A. and Steinberg, M.S. (1981) Two distinct adhesion mechanisms in embryonic chick neural retina cells. I. A kinetic analysis. Dev. Biol. 81:96-105.

Marchase, R.B., Harges, P., and Jakoi, E.R. (1981) Ligatin from embryonic chick neural retina inhibits retinal cell adhesion. Dev. Biol. 86:250-255.

McClain, D.A. and Edelman, G.M. (1982) A neural cell adhesion molecule from human brain. Proc. Natl. Acad. Sci. USA. 79:6380-6384.

Merrell, R., Gottlieb, D.I. and Glaser, L. (1975) Embryonal cell surface recognition. Extraction of an active plasma membrane component. J. Biol. Chem 250:5655-5659.

Roth, S. and Pierce, J.M. (1982) A possible role for the ganglioside GM2 in the development of the avian visual projection, in "Membranes and Genetic Disease", J.R. Sheppard and V. Elving Anderson eds., Alan R. Liss, Inc., New York, p. 165-172.

Rothbard, J.B., Brackenbury, R., Cunningham, B.A. and Edelman, G.M. (1982) Differences in the carbohydrate structures of neural cell-adhesion molecules from adult and embryonic chicken brains. J. Biol. Chem. 257:11064-11069.

Rutishauser, U., Gall, W.E., and Edelman, G.M. (1981) Adhesion among neural cells of the chick embryo. IV. Role of the cell surface molecule CAM in the formation of neurite bundles in cultures of spinal ganglia. J. Cell Biol. 79:382-393.

Rutishauser, U., Grumet, M., and Edelman, G.M. (1983) N-CAM mediates initial interactions between spinal cord neurons and muscle cells in culture. J. Cell Biol. 97:145-152.

Rutishauser, U., Hoffman, S., and Edelman, G.M. (1982) Binding properties of a cell adhesion molecule from neural tissue. Proc. Natl. Acad. Sci. USA 79: 685-689.

Sadoul, R., Hirn, M., Deagostini-Bazin, H., Rougon, G., and Goridis, C. (1983) Adult and embryonic mouse neural cell adhesion molecules have different binding properties. Nature 304: 347-349.

Sperry, R.W. (1963) Chemoaffinity in the orderly growth of nerve fiber patterns and connections. Proc. Natl. Acad. Sci. USA 50: 703-710.

Thiery, J.-P., Brackenbury, R., Rutishauser, U., and Edelman, G.M. (1977) Adhesion among neural cells of the chick embryo. II. Purification and characterization of a cell adhesion molecule from neural retina. J. Biol. Chem. 252: 6841-6845.

Thiery, J.-P., Duband, J.-L., Rutishauser, U., and Edelman, G.M. (1982) Cell adhesion molecules in early chick embryogenesis. Proc. Natl. Acad. Sci. USA 79: 6737-6741.

Thiery, J.-P., Delouvée, A., Gallin, W.J., Cunningham, B.A., and Edelman, G.M. Ontogenetic expression of cell adhesion molecules: L-CAM is found in epithelia derived from the three primary germ layers. Dev. Biol., in press.

Urushihara, H., Ozaki, H.S., and Takeichi, M. (1979) Immunological detection of cell surface components related with aggregation of Chinese hamster and chick embryonic cells. Dev. Biol. 70: 206-216.

PHOSPHOGLUCOSE-CONTAINING GLYCOPROTEINS IN INTRACELLULAR

TRAFFICKING AND NEURONAL RECOGNITION

Richard B. Marchase

Department of Anatomy
Duke University Medical Center
Durham, N.C.

During development of the brain, a neuron's axonal tip may
come into contact with millions of cells with which it could form
synapses. Yet, one of the most striking features of the brain is
the exquisite selectivity and reproducibility of the synaptic
connections that are actually formed.

In his hypothesis of neuronal specificity, Sperry (1943;
1963) attributed this selectivity to interactions during
development between cell-surface molecules on contacting cells.
In the visual system of lower vertebrates, for example, axons of
retinal ganglion cells form a topographically ordered projection
in their synapses with neurons of the optic tectum (Hamdi and
Whitteridge, 1954). Sperry proposed that the axonal tips of each
retinal ganglion cell possessed a unique battery of cytochemical
markers, as did their potential termini in the tectum. It was
suggested that interactions between these molecules resulted in
differential affinities between the individual axons and the
various tectal loci. Sperry reasoned that each axon would adhere
preferentially to, and finally synapse with, an area of tectum
possessing a particular set of surface molecules, while bypassing
sites with less appropriate displays.

Our approach to testing the hypothesis of neuronal
specificity and investigating its biochemical basis began with an
assay capable of directly measuring the adhesion of dissociated
chick neural retina cells to limited areas of tectal surface
(Barbera et al., 1973; Marchase, 1977). Cell bodies from dorsal
retina were shown to adhere preferentially to ventral half-tecta
while cell bodies from ventral retina preferred dorsal
half-tecta, thus mimicking the innervation pattern of the tectum

by the retina. While there is no assurance that the molecular
mechanisms responsible for this in vitro adhesive specificity are
in any way related to those responsible for the in vivo synaptic
specificity, our strategy is to attempt to elucidate the
biochemical basis for the simpler and more accessible in vitro
adhesive selectivity with the goal of ultimately testing these
results for relevance to the more complicated in vivo situation.

One macromolecule that has been identified by us as being
important to neuronal interactions is a filamentous cell-surface
protein called ligatin (Jakoi and Marchase, 1979; Marchase et
al., 1981). Our evidence suggests that at least some of the
glycoproteins participating in recognition do not directly
interact with the lipid bilayers of plasma membranes but rather
are secured to the cell surface through their binding to ligatin,
which functions generally as a baseplate for the attachment of
peripheral glycoproteins to the membrane (Jakoi et al., 1976;
Marchase et al., 1982a).

We have recently determined the mechanism for the attachment
of such peripheral glycoproteins to ligatin and thus to the cell
surface (Marchase et al., 1982a). The glycoproteins all bear
specially derivatized oligosaccharides that are recognized and
bound by ligatin. These high mannose-type oligosaccharides are
unique due to the presence of terminal glucose residues linked to
the oligosaccharide via phosphodiester bonds.

Our finding that proteins recognized by ligatin all contain
oligosaccharides containing phosphodiester-linked glucose has
prompted our consideration of a related but independent problem
in neurobiology, the mechanism by which newly synthesized
proteins are sorted to their appropriate subcellular
destinations. The relevance of ligatin to problems of
intercellular adhesion and our approach to the problems of
intracellular trafficking will be reviewed in this chapter.

LIGATIN FROM THE ILEUM OF SUCKLING RATS

In suckling neonates epithelial cells of the ileum are
characterized by a highly invaginated, interconnecting system of
plasma membrane beneath the microvilli, the endocytic complex.
In electron microscopic studies (Porter, et al., 1967; Wissig and
Graney, 1968) the lumenal surfaces of these membranes were found
to be partially covered by two-dimensional arrays of 7.5 nm
particles. Knutton et al. (1974) suggested that these arrays
were composed of aligned rows of the particles, since during the
preparation of these membranes individual filaments with a
periodic unidimensional structure were dissociated from the
membrane surface.

Jakoi et al. (1976) pursued this observation and found that
extraction of these filaments was enhanced by treatment of the
membranes with 10 mM $CaCl_2$. In addition, they showed that
dialysis against EGTA, a calcium chelator, resulted in
depolymerization of the filaments. Molecular sieve chromatography separated the constituent proteins into two
populations. The excluded peak contained β-N-acetylhexosaminidase activity and was in part morphologically similar to the
7.5 nm particles that appeared as rows on the intact ileal
membranes. Eluting later from the sizing column with an apparent
molecular weight of 20,000 daltons or less was a second peak of
protein which, upon addition of 20 mM $CaCl_2$, formed 3 nm
filaments of varying lengths. The authors suggested that these
filaments formed from the calcium-dependent polymerization of the
smaller protein and acted as attachment sites for the 7.5 nm
β-N-acetylhexosaminidase particles to produce the decorated
filaments seen on the membranes. This model provided an
explanation for the regular arrays of these particles seen in
electron micrographs. It was also consistent with the
morphological observation that the particles seemed to be only
peripherally associated with the lumenal unit membrane. A
schematic of this model is provided below. The filament-forming
protein was named ligatin, from the Latin _ligare_ which translates
"to bind together".

LIGATIN FROM EMBRYONIC CHICK NEURAL RETINA

Some of the properties of ileal ligatin, e.g., its plasma
membrane location and its relatively small molecular weight are
identical to characteristics of an inhibitor of retinal adhesion
characterized by Merrell et al. (1975). These and other physical
similarities led to an investigation in retina for a ligatin-like
molecule (Jakoi and Marchase, 1979).

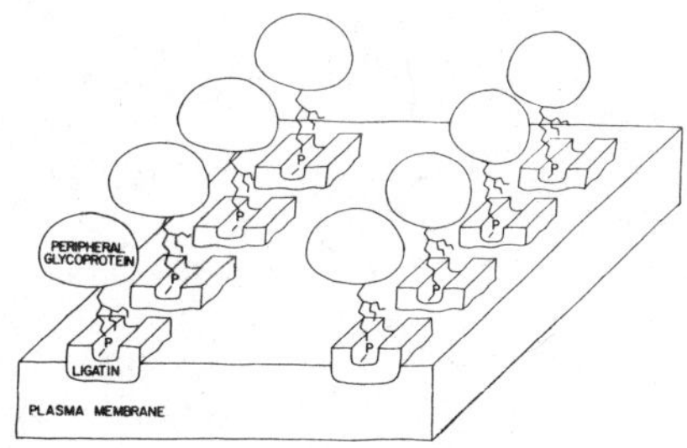

Figure 1. A schematic showing peripheral glycoproteins bound to
the cell surface baseplate ligatin.

Retinal plasma membranes were examined by electron
microscopy using negative staining techniques. Most of the
membranes were covered in part with randomly organized filaments
4.5 nm in diameter. When the plasma membranes were treated with
40 mM Ca^{++}, individual filaments 4.5 nm in diameter were seen
free of the membranes scattered over the grids. After dialysis
against EGTA, pH 8.0, filamentous materials were no longer seen.
The release of filamentous material from plasma membrane surfaces
after treatment with Ca^{++} and the subsequent disappearance of the
filaments after dialysis against EGTA are identical to results
obtained with ligatin-hexosaminidase filaments isolated from
ileum. However, both the membrane-bound filaments and the free
filaments differ in morphology from those seen in ileal
preparations.

Subsequent molecular sieve chromatography of the retinal
extracts yielded two protein peaks, a heterogeneous peak eluting
first from the sizing column and a second protein chromatographin
with an apparent molecular weight of approximately 10,000 daltons
(Jakoi and Marchase, 1979). Unlike the preparations from ileum
the retinal proteins eluting early from the sizing column were
found to have minimal β-N-acetylhexosaminidase activity. They
also bore no morphological resemblance to the ileal enzyme. In
contrast to these results, the smaller protein seemed very
similar to its analog from ileum. After dialysis against 5 mM
CaCl$_2$ negative stain electron microscopy of the protein which
previously exhibited a molecular weight of 10,000 daltons
revealed polymerization of the protein into filaments
approximately 3 nm in diameter. These filaments were identical
in morphology to the repolymerized filaments of ileal ligatin.
SDS-polyacrylamide gel electrophoresis confirmed an apparent
molecular weight of approximately 10,000 daltons. An amino acid
composition performed on retinal ligatin revealed a relatively
polar profile (Jakoi and Marchase, 1979), and was similar to the
composition subsequently obtained for the ileal protein (Gaston
et al., 1982).

Jakoi and Marchase (1979) proposed that the retinal protein
also be considered ligatin. Because of the electron microscopic
findings in ileum, ligatin in neural retina was suggested to
serve as a baseplate on the external surface of the cells for the
attachment of other, more peripheral glycoproteins. The finding
that retinal ligatin, following its separation by molecular sieve
chromatography, repolymerized into 3nm filaments but that native
filaments from plasma membranes of retina had diameters of 4.5 nm
suggested that the membrane filaments were decorated with
additional proteins. Both electron microscopic and biochemical
observations, however, suggested that the glycoproteins bound to

164

the ligatin filaments in retina differed from the 7.5 nm
particles of β-N-acetylhexosaminidase found decorating ileal
ligatin.

As was mentioned earlier, a primary stimulus for seeking
ligatin in neural retina was the similarity between ileal ligatin
and the proteinacious inhibitor of retinal cell reaggregation
described by Merrell et al. (1975). Since ileal ligatin seemed
to function as a baseplate for the attachment of external cell
surface proteins, it seemed possible that an analogous function
by a similar protein in retina could be responsible for the
inhibitory activity: if proteins that mediated intercellular
adhesion were bound to retinal plasma membranes via ligatin, then
exogenous soluble ligatin might compete for this material and
thus inhibit adhesion rates.

In order to ascertain whether retinal ligatin possessed
biological activity similar to the preparation of Merrell et al.
(1975), single cell reassociation studies were carried out in the
absence and presence of retinal ligatin (Marchase et al., 1981).
In a typical experiment, 70% of the single cells present at time
0 became associated with other cells by 30 min of incubation in
the absence of ligatin. When ligatin at 20 µg/ml was added to
the assay only 20% of the single cells reassociated. This
represented a 70% inhibition of aggregation.

Further experiments addressing a role for ligatin in
reaggregation attempted to distinguish between two separable
adhesion mechanisms that have been described in neural retina and
other cells. One is active independent of the presence of Ca^{++}
ions, and in retina involves the integral membrane protein called
CAM (Edelman, 1983). The other is active only when Ca^{++} is
present (Urishihara et al., 1976; Thomas et al., 1981; Balsamo
and Lilien, 1982). The experiments of interest here utilized
αglucose-1-phosphate (αGlc-1-P). When intact retinal cells
labeled metabolically with ^{32}P were treated with 10 mM αGlc-1-P
glycoproteins containing oligosaccharide-associated ^{32}P were
solubilized (Marchase et al., 1982b). This treatment did not
adversely affect the cells' viability nor their capacity for
adhesion in the absence of Ca^{++}. However, the addition of Ca^{++}
to reaggregating control cells resulted in an approximate
two-fold stimulation of adhesion, consistent with the
activitation of a Ca^{++}-dependent adhesion mechanism (Urushihara
et al., 1976). In the αGlc-1-P-treated cells this Ca^{++}
stimulation did not occur (Marchase et al., 1982b). This may
indicate that αGlc-1-P was effective in removing proteins from
the cell surface that are active in Ca^{++}-dependent adhesion. The
relation of αGlc-1-P to ligatin is discussed in the next section.

Consistent with this suggestion is a recent report by
Balsamo and Lilien (1982). They reported that a molecule
implicated by them in Ca^{++}-dependent adhesion can exist in
multiple forms, separable upon sucrose density gradient
fractionation. They found that treatment of their preparation
with 0.5 mM αGlc-1-P caused a shift in the activity away from the
heavier form to a lighter form. They suggested that this effect
could be due to the dissociation of the adhesion molecules from a
complex requiring binding to ligatin for its integrity.

AFFINITY CHROMATOGRAPHY TO IMMOBILIZED LIGATIN

Electron microscopic observations (Jakoi et al., 1976) led
to the proposal that ligatin functions as a plasma membrane
baseplate for the attachment of other, more peripheral
glycoproteins to the external cell surface. The association that
occurs between ligatin and these more peripheral glycoproteins
has since been investigated utilizing ligatin covalently attached
to inert matrices. In experiments with neural retina ^{32}P was
used to metabolically label the glycoproteins which
co-solubilized with ligatin; their binding to a ligatin affinity
column was then monitored radioactively (Marchase et al., 1982a).

When proteins that had co-solubilized with ligatin from
neural retina were passed over an Affi-gel 10 column covalently
coupled with 700 µg of ligatin approximately 25% of the ^{32}P
applied to the column was retained (Marchase et al., 1982a). The
bound radioactivity was resistant to removal by neutral sugars
but was almost completely eluted with 10 mM αGlc-1-P.
α-Glucose-1-phosphate was by far the most effective of the
hexoses phosphorylated at the 1-carbon, while βGlc-1-P was
ineffective. Among the sugars phosphorylated at carbon 6,
Man-6-P was most effective, although it required higher
concentrations than αGlc-1-P for comparable levels of elution.
The efficacy of αGlc-1-P in eluting retained proteins from the
ligatin affinity column is consistent with the presence of
oligosaccharides on the bound proteins that contain glucose
residues attached via phosphodiester bonds through their
1-carbons and which mediate the attachment of the glycoproteins
to ligatin. The efficacy of Man-6-P suggests that the
phosphodiester bonds could link the terminal glucose residues to
the 6-carbons of penultimate mannose residues.

^{32}P-containing oligosaccharide chains were isolated from the
glycoproteins following treatment with endoglycosidase H (endo
H). The substrate specificity of this enzyme (Tarentino and
Maley, 1974) suggests that the oligosaccharide chains mediating
this binding were of the high mannose-type. This procedure

resulted in deglycosylated proteins which would no longer bind to
the affinity column. However, the isolated ^{32}P-labeled
oligosaccharides displayed affinity for the column (Marchase et
al., 1982a). Experiments with the cation exchange resin QAE
Sephadex suggested that approximately 25% of the endo H-sensitive
^{32}P was present as phosphodiester bonds.

Mannose-6-phosphate was also detected in complete acid
hydrolysates of the oligosaccharide using linked enzymatic assays
(Marchase et al., 1982a). The identification of glucose as the
sugar distal to the phosphate has been confirmed directly by the
appearance of glucose-1-phosphate following alkaline hydrolysis
of these oligosaccharides (unpublished data). The alkaline
hydrolysis seems to take place through the formation of
glucose-1,2 cyclic phosphate (Paladini and Leloir, 1951), and
supports the suggestion from affinity column studies that glucose
is present in the α configuration.

SYNTHESIS OF THE PHOSPHO-OLIGOSACCHARIDES FROM NEURAL RETINA

The synthetic origin of the phosphate present in the
ligatin-associated oligosaccharide has been investigated by Koro
and Marchase (1982). Similar to the results found by Reitman and
Kornfeld (1981) and Hasilik et al. (1981) in their examinations
of lysosomal enzyme phosphorylated oligosaccharides, the
presence of phosphate in the ligatin-associated oligosaccharides
seems to occur due to the intact transfer of a sugar-1-phosphate
by a phosphoglycosyltransferase.

The reaction observed by Koro and Marchase (1982) for the
UDP-Glc:glycoprotein glucose-1-phosphotransferase (GlcPTase) in
the retinal system is:

UDP-Glc + Oligosaccharide → Oligosaccharide-P-Glc + UMP

This was detected in homogenates from 10 day embryonic chick
neural retina using doubly-labeled UD^{32}P-^{3}H-glucose. ^{32}P was
incorporated solely into endo H-sensitive oligosaccharides, and
was not affected by inclusion in the assays of excesses of
unlabeled breakdown products or potential contaminants. Excess
unlabeled UDP-Glc, however, completely abolished the ^{32}P
incorporation. The reaction had a pH optimum of 6.2 and was
stimulated by manganese, with 5 mM being the optimal
concentration. The major acceptors of ^{32}P from UD^{32}P-Glc in
neural retina are distinct from the primary acceptors for
UD^{32}P-GlcNAc (Koro and Marchase, in preparation).

A POSSIBLE ROLE FOR THE GLCPTASE IN INTRACELLULAR TRAFFICKING

The phosphodiester linkage described for ligatin-associated
retinal cell-surface proteins does not seem to undergo cleavage
in vivo to expose Man-6-P, since phosphodiester-linked glucose is
present in cell-surface preparations and may be required for
binding to ligatin (Marchase et al., 1982). This raises the
possibility that the phosphodiester-linked glucose could be a
distinct subcellular trafficking signal from the Man-6-P signal
active in lysosomal localization (Sly and Fisher, 1982; Marchase
et al., 1984). A diagrammatic summary of this proposal is
presented below.

Three classes of polypeptides manufactured on
membrane-associated ribosomes and then released into the lumen of
the endoplasmic reticulum are depicted: ⚐ , those destined for
secretion; ⚭ , those destined for lysosomes; ⬡ , those destined
for association with ligatin at the cell surface. These classes
of proteins are shown to bear "core" oligosaccharides produced in
the ER and the cis regions of the Golgi. In the absence of a
diverting signal, the first class of polypeptides traverse the
Golgi and are localized to secretory vesicles. The second and
third classes are marked via the acquisition of phosphorylated
sugars for segregation to other vesicles, and for eventual
localization to specific subcellular destinations. Most of the
30 to 40 glycoproteins destined for localization to lysosomes
have been suggested to possess a common characteristic in their
tertiary structures (Reitman and Kornfeld, 1981b) that codes for
their recognition by a GlcNAcphosphotransferase. This enzyme

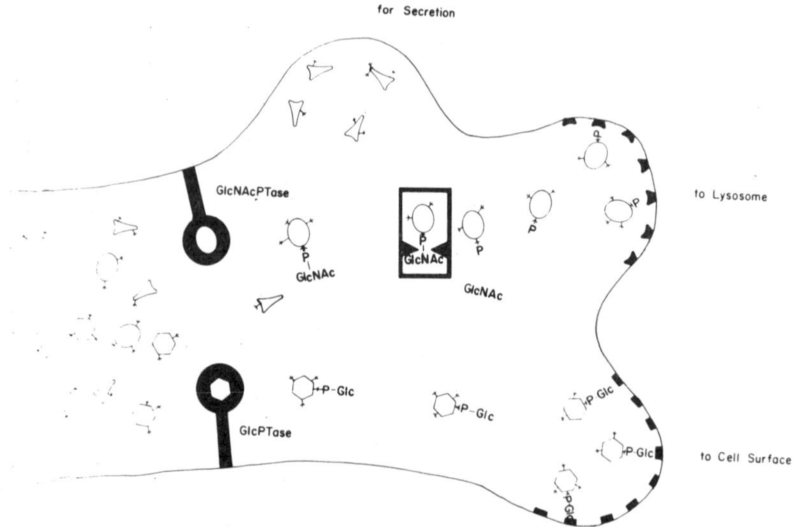

Figure 2. The possible involvement of phosphorylated oligosaccharides in protein trafficking.

adds phosphodiester-linked GlcNAc to the high mannose-type
oligosaccharides on these proteins. The derivatized proteins are
then further processed by the enzymatic removal of the terminal
GlcNAc residues, exposing the Man-6-P's that are recognized by
the Man-6-P receptor. This recognition ultimately leads to the
segregation of these proteins in vesicles destined for primary
lysosomes (for review, see Sly and Fisher, 1982).

We propose a parallel mechanism that could segregate another
class of glycoproteins from those just mentioned and from those
destined for secretion. We suggest these glycoproteins
(   ) could bear a characteristic in their tertiary
structures that would code for their recognition by the GlcPTase.
The phosphodiester-linked glucose would then be recognized by an
intracellular receptor distinct from that utilized in the
lysosomal localization scheme. This receptor could be ligatin or
another protein of similar binding specificity. This recognition
would ultimately lead to the segregation of these proteins in
vesicles destined for the cell surface. Once at the cell surface
these proteins would establish a stable presence there through
association with the cell surface baseplate ligatin.

Alternatively, the phosphodiester-linked glucose-containing
proteins could be handled by the cell in the same manner as
secretory proteins. It is possible that both the
ligatin-associated glycoproteins and those which are to be
secreted (   ) could travel together to the cell surface.
Once the vesicles fused and opened there, the appropriately
derivatized glycoproteins could form or maintain an association
with ligatin while the secretory proteins would diffuse away.

LOCALIZATION OF THE PHOSPHOGLUCOSYLTRANSFERASE REACTION AND
AXOPLASMIC TRANSPORT

The visualization of incorporated ^{32}P-Glc1P would not only
help substantiate the role we have hypothesized for GlcPTase in
intracellular trafficking, but might also provide the means to
follow a specific subclass of proteins via intracellular vesicles
to their ultimate cell-surface location.

This problem is complicated by the well-established finding
that nucleotide sugars do not penetrate cells' plasma membranes.
Hiller and Marchase (in preparation) have utilized intracellular
pressure injection to deliver (β^{32}P)UDP-Glc to identified giant
neurons within the _Aplysia_ abdominal ganglia. Results to date
have been extremely encouraging. We have been able to inject from
5,000 to 100,000 cmp's into individual neurons, and have
demonstrated sufficient levels of incorporation into glycoproteins
to allow biochemical analyses to proceed. The oligosaccharide

products appear biochemically to be similar to those described in neural retina; however, attempts at light-level autoradiography of these macromolecules have been disappointing, in part due to the energy of the ^{32}P emission.

One extremely interesting result has emerged. Along the length of the axon of the injected cell ^{32}P, exclusively associated with macromolecules, was detected. This result suggests to us that vesicles carrying the phosphoglucose-terminating glycoproteins may not solely be destined for delivery locally to the cell surface, but may also be coded for transport down the length of an axon for eventual incorporation at a synaptic ending.

PROTEINS CAPABLE OF BINDING αGLC-1-P-6-MAN

In investigations designed to purify any cellular proteins which might act as receptors for the phospho-oligosaccharide synthesized by the GlcPTase, Hiller et al. (1984) have utilized an artificial ligand which is analogous to this structure. The ligand, α-glucose-1-P-6-mannose-phenylamine, was coupled to cyanogen bromide-activated Sepharose and used in affinity chromatography studies. Two results merit special mention. First, crude membrane preparations from either suckling pig ileum or embryonic chick brain were extracted with the chaotropic agent lithium diiodosalicylate (LIS). The resulting supernatant was passed over the affinity column described above. Greater than 95% of the protein ran directly through the column. However, elution with high concentrations of NaCl released proteins which had been retained. Polyacrylamide gel electrophoresis of these proteins from either source revealed only three bands, ranging in size from 10,000 to 17,000 daltons. Secondly, plasma membrane-enriched fractions from these same sources were extracted with $CaCl_2$ as described in the first two sections. The protein peak which eluted late from molecular sizing columns and has been identified with ligatin was then applied to the αGlc-1-P column. Nearly 100% of the protein bound to the column, and after its release, electrophoresed identically to the retained proteins from the LIS extract. Other purified proteins, for example bovine serum albumin, showed no affinity for the column.

The finding of three low-molecular weight peptides capable of binding to this phospho-oligosaccharide is intriguing. The manner in which the forms of the binding protein relate to one another and the subcellular distributions of each are currently under investigation.

FUTURE DIRECTIONS

The neuron shares with all cell types the necessity for
selectively distributing newly synthesized proteins to the various
organelles. Beyond this, the neuron possesses an expansive and
regionally specialized plasma membrane which, as discussed in a
recent review by Hammerschlag (1983), may require additional
mechanisms for delivering distinct macromolecules to specific
microdomains of axonal, somatic, or dendritic membranes.

Certainly some of the proteins which participate in the
processes of interneuronal recognition would be expected to be
axonally transported to the growth cone region of the cell. The
possible involvement of ligatin and its associated
phosphoglycoproteins in adhesion and the preliminary finding that
proteins labeled by the GlcPTase are axonally transported suggest
that the model for trafficking discussed in this chapter might
provide one of the mechanisms involved in these processes.

ACKNOWLEDGMENTS

I thank Lillian Koro, Ann Hiller and Ronald Myatich for
helpful comments. I also thank Rachel Fink for the illustrations
and am deeply indebted to Barbara McPartland for her patience and
excellent secretarial support. Work from this laboratory is
supported by NIH EY 04480, NS 06233, GM 31381, and by a Nanaline
H. Duke Scholarship.

REFERENCES

Balsamo, J. and Lilien, J. 1982, An N-acetylgalactosaminyl-
     transferase and its acceptor in embryo chich neural
     retina exist in interconvertible particulate forms
     depending on their cellular location, J. Biol. Chem.,
     257:349-354.
Barbera, A.J., Marchase, R.B., Roth, S., 1973, Adhesive
     recognition and retinotectal specificity,
     Proc. Natl. Acad. Sci. USA, 70:2482-2486.
Edelman, G.M., 1983, Cell adhesion molecules, Science 219:
     450-457.
Gaston, S.M., Marchase, R.B. and Jakoi E.R., 1982, Brain
     ligatin: a membrane lectin that binds
     acetylcholinesterase, J. Cell. Biochem., 18:447-459.
Hamdi, F.A., Whitteridge, E., 1954, The representation of the
     retina on the optic lobe of the pigeon, Q. J. Exp.
     Physiol., 39:111-119.
Hammerschlag, R., 1983, How do neuronal proteins know where they
     are going?...speculations on the role of molecular
     address markers, Dev. Neurosci., 6:2-17.
Hasilik, A., Klein, U., Waheed, A., Strecker, G. and von Figura,
     K., 1980, Phosphorylated oligosaccharides in lysosomal
     enzymes: identification of
     α-N-acetylglucosamine(1)phospho(6)mannose diester
     groups, Proc. Natl. Acad. Sci. USA 77:7074-7078.
Hiller, A.M., Hindsgaul, O., Myatich, R., and Marchase, R.B.,
     1984, αGlucose-1-phosphate binding proteins isolated
     by affinity chromatography, submitted.
Jakoi, E.R. and Marchase, R.B. (1979). Ligatin from embryonic
     chick neural retina. J. Cell Biol., 80:642-650.
Jakoi, E.R., Zamphghi, G. and Robertson, J.D., 1976, Regular
     structures in unit membranes: II. Morpholocial and
     biochemical characterization of two water-doluble
     membrane proteins isolated from the suckling rat ileum.
     J. Cell Biol., 70:97-111.
Knutton, S., Limbrick, A.R. and Robertson, J.D., 1974, Regular
     structures in membranes. I. membranes in the endocytic
     complex of ileal epithelial cells. J. Cell. Biol., 62:
     679-694.
Koro, L.A., and Marchase, R.B., 1982, A UDP-glucose:glycoprotein
     glucose-1-phosphotransferase in embryonic chicken neural
     retina, Cell, 31: 739-748.
Marchase, R.B., 1977, Biochemical investigations of retinotectal
     adhesive specificity, J. Cell Biol., 75:237-257.
Marchase, R.B., Harges, P. and Jakoi, E.R., 1981, Ligatin
     from embryonic chick neural retina inhibits retinal cell
     adhesion, Developmental Biol., 86:250-255.
Marchase, R.B., Koro, L.A. and Hiller, A.M., Receptors
     for Glycoproteins with Phosphorylated Oligosaccharides.

in "The Receptors," Vol II P.M. Conn, ed., Academic
Press, New York (1984).

Marchase, R.B., Koro, L.A., Kelly, C.M and McClay, D.R., 1982,
Retinal ligatin recognizes glycoproteins bearing
oligosaccharides terminating in phosphodiester-linked
glucose. Cell 28, 813-820.

Marchase, R.B., Koro, L.A., Kelly, C.M. and McClay, D.R., 1982b,
A possible role for ligatin and the phosphoglycoproteins
it binds in calcium-dependent retinal cell adhesion,
J. Cell Biochem., 18:461-468.

Merrell, R., Gottleib, D.I. and Glaser, L., 1975, Embryonal cell
surface recognition. Extraction of an active plasma
membrane component, J. Biol. Chem., 250:5655-5659.

Paladini, A.C. and Leloir, L.T., 1952, Studies on uridine-diphos-
phate-Glucose, Biochem. J., 51:426.

Porter, K.R., Kenyon, K. and Badenhauser, S., 1967,
Specializations of the unit membrane, Protoplasma, 63:
262-274.

Reitman, M.L. and Kornfeld, S., 1981a, UDP-N-Acetylgluco-
samine:glycoprotein N-acetylglucosamine-1-phospho-
transferase. Proposed enzyme for the phosphorylation
of the high mannose oligosaccharide units of lysosomal enzymes, J. Biol. Chem., 256: 4257-4281.

Reitman, M.L. and Kornfeld, S., 1981b, Lysosomal enzyme
targeting, J. Biol. Chem., 256:11977-11980.

Sly, W.S. and Fisher, H.D., 1982, The phosphomannosyl
recognition system for intracellular and intercellular
transport of lysosomal enzymes, J. Cell Biochem.,
18:531-549.

Sperry, R.S., 1943, Visuomotor coordination in the newt
(Triturus viridescens) after regeneration of the optic
nerves, J. Comp Neurol., 79:33-35.

Sperry, R.W., 1963, Chemoaffinity in the orderly growth of nerve
fiber patterns and connections, Proc. Natl. Acad. Sci.
USA, 50:703-710.

Tarentino, A. and Maley, F., 1974, Purification and properties
of an endo-β-N-acetylglucosaminidase from Streptomyces
griseus, J. Biol. Chem. 249:811-817.

Thomas, W.A., Thompson, J., Magnani, J.L. and Steinberg, M.S.,
1981, Two distinct adhesion mechanisms in embryonic
neural retina cells, Develop. Biol., 81 379-385.

Urushihara, H., Takeichi, M., Hakura, A. and Okada, T.S., 1976,
Different cation requirements for aggregation of BHK
cells and their transformed derivatives, J. Cell Sci.,
22: 685-695.

Wissig, S.L. and Graney, D.O., 1968, Membrane modifications in
the apical endocytic complex of ileal cells,
J. Cell Biol., 39:564-579.

NEW INSTRUMENTATION FACILITATES THE STUDY OF GENES CODING

FOR MOLECULES INVOLVED IN CELL SURFACE RECOGNITION

William J. Dreyer, Janet Roman and David B. Teplow

Division of Biology
California Institute of Technology
Pasadena, California 91125

INTRODUCTION

During the last two decades studies of the molecular mechanisms by which antibody diversity is generated have provided us
with fundamental insights into the structure, organization and
programmed movement of immunoglobulin genes during development
(Potter et al., 1964; Dreyer et al., 1967; Hood et al., 1975;
Huang and Dreyer; 1978; Leder, 1982).   Progress in this
specialized area of developmental biology has occurred far more
rapidly than in any other developmental system. (An appreciation
of the level of knowledge extant in 1967 may be obtained by
reading the proceedings of the Cold Spring Harbor Symposium on
Quantitative Biology, volume 32, 1967.) The depth of genetic and
molecular understanding which has been gained in this field is due
in large part to the existence of myelomas, tumors of antibody-
producing cells, from which gram quantities of antibody molecules
could easily be obtained. These amounts of protein were necessary
and sufficient, given the methods and instrumentation available in
the sixties, to do structural analyses and to obtain protein
sequence information about these molecules.   It was the
collection, compilation and analysis of these data which led to
the discovery that programmed movement of genetic material plays a
key role in generating diversity in cells of the developing immune
system (Dreyer and Bennett, 1965).

It has been postulated that a similar mechanism may be
involved in the generation of other families of cell surface
proteins which function in the highly specific cellular migrations
and associations which occur during embryological development
(Dreyer et al., 1967; Hood et al., 1977; Dreyer, 1984). Many

scientists believe that cells in the developing embryo sense their environment by means of cell surface receptors which appear on specific lineages of cells during development. Cell surface receptors are generally present in extremely small amounts. Although there was great interest in these molecules in the early seventies, studies of such low abundance proteins, and of their corresponding mRNAs and genes, were not feasible considering the methods and instruments available at the time. It was for this reason that one of us (WJD), in collaboration with Lee Hood and others at Caltech and the Jet Propulsion Laboratory (JPL) in Pasadena, conceived and began development of sensitive new instruments and techniques to study these and other biologically important molecules. We would like to present here a brief review of the progress that has been made over the last decade in the development of a microchemical facility at Caltech and in its current application by our laboratory in the study of cell recognition molecules and the genes which code for them.

MICROCHEMICAL INSTRUMENTATION

Gas-Liquid Solid-Phase Protein Sequenator

The first instrument developed for the microchemical facility was an ultrasensitive protein sequenator (Dreyer, 1977; Hewick et al., 1981). This instrument automatically performs the stepwise removal of amino acids from polypeptides using chemistry extensively developed by Pehr Edman (for a review, see Braunitzer, 1977). It is capable of providing extended sequence information on picomole (microgram) levels of proteins and peptides. This represents greater than a 10,000-fold increase in sensitivity compared with the first automated sequenator developed by Edman and Begg (1967). The major features responsible for this increase in performance are the noncovalent immobilization of the protein sample on a glass matrix impregnated with the quaternary polymeric ammonium compound polybrene, the use of gaseous reagents at steps during which the sample would be soluble in liquid reagents, and the miniaturization of all hardware components of the instrument. These changes in design result in better retention of the sample during sequencing, more efficient reagent delivery and removal and generation of fewer side products of the chemistry which can interfere with subsequent analysis of the phenylthiohydantoin (PTH) derivatives of the Edman reaction. In practical terms, it makes possible the structural study of proteins which may be present at levels of only 100 molecules/cell. Studies of proteins expressed in lower amounts are hindered primarily by our inability to purify such minute quantities of material, using classical biochemical methods, and by the sensitivity of our current high performance liquid chromatography (HPLC) system, which has a prac- tical limit of detection of approximately a picomole.

176

Miniaturized Mass Spectrometer

To provide the increased sensitivity required for PTH-amino acid analysis in the femtomole range a miniaturized mass spectrometer, which is sensitive enough to detect individual molecular ions, has been constructed through a collaborative effort of scientists at Caltech and JPL (Dreyer et al., 1974). The instrument utilizes an electro-optical detector array in which each molecular ion induces the formation of approximately 10^6 photons. These photons impinge on a phosphorescent screen creating an image which is quantified through a molecular weight range of 25-500 amu by 5120 individual photodiode-capacitor pairs. By measuring the charge stored in each capacitor and interfacing this information with a computer, one can obtain the complete mass spectrum of a sample in minutes. In the future we expect to interface the mass spectrometer simultaneously with a number of sequenators so as to automatically perform the analysis of the resulting PTH amino acids. We anticipate this system reducing our analysis time by a factor of ten and the amount of protein required to obtain a sequence by a factor of a thousand.

Protein Synthesizer

The protein synthesizer incorporates solid phase peptide chemistry (see reviews by Barany and Merrifield, 1980, and by Kent, 1980) in the automated synthesis of polypeptides containing up to 70 residues. This extraordinary chain length is possible due to the 99.8% addition efficiency of the instrument. Among its many uses, the protein synthesizer is used for construction of partial and/or complete proteins for structure/function studies, for production of biologically active peptides, including hormones and growth factors, and for production of synthetic vaccines. In addition, synthetic peptides are of use in the production of mono-clonal antibodies and as reagents in a variety of immunochemical, enzymatic and biochemical assays.

DNA Synthesizer

To isolate the structural gene for a protein, we synthesize oligodeoxynucleotides, label them with ^{32}P and use them to probe cDNA and/or genomic libraries (see below). The sequences of these oligodeoxynucleotides are derived by using the genetic code to reverse translate 4-7 amino acid long sequences, obtained from the gas phase sequenator. The protein sequences chosen are those for which the total number of possible DNA coding sequences is the smallest. All of these coding sequences are then synthesized simultaneously as an equimolar mixture on the DNA synthesizer. Besides their use in cloning, we utilize the oligodeoxynucleotides in studies of gene structure using Southern and Northern blotting techniques. The instrument uses miniaturized hardware based on that of the protein sequenator and incorporates phosphoramidite chemistry in the solid phase synthesis of the DNA (Horvath et al., 1984). Cycle times of 20-30 minutes allow up to 50 cycles/day to be accomplished. Besides the synthesis of probes, the DNA synthesizer can be used for the construction of partial or complete genes, both normal and incorporating specific predetermined mutations, as well as for construction of linkers and other sequences useful in gene cloning.

Computerization in the Microchemical Facility

Computer technology is being applied in two main areas in the facility: (1) user friendly control of the operation of the instrumentation; and (2) data analysis and manipulation. Each instrument is designed to be operated by microprocessor based controllers. This miniaturizes further the hardware in the system and lends great programming flexibility to the instrumentation. In both protein and DNA systems, sequence dependent changes in chemistry are crucial for maximizing yields. The microprocessor based controller allows one to easily and rapidly design unique programs for every sample to be dealt with. In addition, it has the ability to continuously check for errors in the operating parameters of the system and to take necessary actions if errors are detected.

Data obtained from protein sequencing can be entered directly into the facility's computer using analog-digital converters which interface with the UV detector of the HPLC system. A variety of graphics programs exist to plot the data in various formats. Programs also may be utilized to calculate repetitive yields and perform other statistical manipulations. One of the most useful aspects of the computer system is its ability to create sequence files and to manipulate them. Homology comparisons can be done among protein or DNA sequences using internal data bases or on-line access to national sequence data bases. For example, one such search, done in this case by Russ Doolittle at the University of California at San Diego, suggested that a human melanoma cell surface "tumor antigen" was related to a family of iron transport proteins. Subsequent experiments confirmed this very interesting relationship and shed light on the function of that particular tumor-associated protein (Brown et al., 1982).

MICROCHEMICAL STUDIES OF CELL SURFACE PROTEINS INVOLVED IN CELL-CELL INTERACTIONS AND NEUROLOGICAL DEVELOPMENT

Monoclonal antibodies have been produced which define antigens that are expressed only on subsets of leukocytes (Milstein and Lennox, 1980) or subsets of neurons or glia (reviewed by de Blas, 1984) and whose appearances are under extremely precise temporal control. Molecules with these characteristics may be involved in the cellular interactions occurring in the immune system and in development of the nervous system. In our collaborative studies, proteins have been purified from detergent extracts of leukocytes or brain by affinity chromatography on columns to which the relevant monoclonal antibody had been covalently attached. The eluates from these columns have then been purified further either by sodium dodecyl sulfate polyacrylamide gel electrophoresis (SDS-PAGE) or by size exclusion chromatography (SEC) using Sephadex gels. Proteins from SDS gels are visualized with Coomassie blue, excised, electroeluted, dialyzed and concentrated (Hunkapiller, 1983). Proteins purified by SEC are dialyzed and concentrated. At this stage the proteins are pure as determined by analytical SDS-PAGE and silver staining. Although not utilized in the studies to be presented here, it is important to note that ion exchange, reversed phase and size exclusion HPLC methods have also been of great utility in the preparation of proteins and peptides for sequencing.

Cell-Cell Interactions in the Immune System

Immune responses are elicited through the cooperative interactions of a number of specific cell types (for a review see Katz, 1977). Cells of each type possess unique constellations of cell surface proteins which mediate these interactions. One family of proteins that is known to play a role in cell-cell interactions is composed of non-covalently associated heterodimers of molecular weights 95,000 and 170,000-180,000 (Trowbridge and Omary, 1981; Kürzinger and Springer, 1982). In collaboration with Timothy Springer's group at the Dana-Farber Cancer Institute, we have studied two members of this family: (1) LFA-1, a heterodimer associated with lymphocytes (Kürzinger et al., 1981); and (2) Mac-1, a heterodimer associated with macrophages and other phago-cytic cells (Springer et al., 1979). Amino terminal protein sequences were obtained from the high molecular weight chains of each molecule after affinity chromatography and preparative gel electrophoresis. The two chains display a significant degree of homology, which one would predict based on the molecular weights and functional similarities of the polypeptide pairs. We have used the sequence information from the Mac-1 heavy chain to design two separate pools of oligodeoxynucleotide probes, each probe being twenty nucleotides in length. These have been synthesized and used to isolate two genomic clones which hybridized with both sets of probes. By screening the library with two pools of probes representing non-overlapping regions of the protein we were able to eliminate from consideration more than fifty clones which were positive with only one of the pools. We chose to directly screen a genomic library instead of a cDNA library to circumvent the problems associated with producing full-length cDNA copies of genes as large as Mac-1. By sequencing the two positive clones, deriving the corresponding protein sequence, and comparing that with the amino acid sequence obtained from our study of the polypeptide, we can determine if we have indeed isolated the Mac-1 structural gene. We will then be in a position to elucidate the structure and organization of this gene and to use portions of it as probes to isolate other members of this family. If desired we will also be able to determine the complete DNA and protein sequences of members of this family and will be able to follow the expression and movement (if any) of the genes during development. We are particularly interested in learning whether the genes which code for various members of this family of cell surface receptors are tandemly arrayed on the same chromosome and whether the sequences of DNA control regions of each gene provide insights into the mechanism(s) of differential gene expression in cells of these lineages, all of which are derived from a common hemopoietic stem cell.

Development of the Drosophila Retina

Fujita et al. (1983) produced a panel of 146 monoclonal antibodies (MAbs) specific for the Drosophila melanogaster nervous system. One of these, MAb24B10, defined a glycoprotein antigen with an apparent molecular weight (Mr) of 160,000 which was expressed exclusively on photoreceptor cells and their axons. Antigen 24B10 (Ag24B10) first appears on retinal neurons late in the third larval instar at the time photoreceptor cell determination occurs. It appears to be the earliest marker for these cells. To study further its structure and the structure and regulation of its gene and mRNA transcripts we sequenced a portion of the amino terminus of the protein and designed three sets of oligodeoxynucleotide probes to be used by the Benzer group in their molecular genetic studies (Zipursky et al., 1984). These probes were synthesized and used to screen a Drosophila genomic library constructed in phage λ (Maniatis et al., 1978). Positive clones were purified, rescreened and subcloned. One such clone was shown to contain the sequence coding for the amino terminus of Ag24B10 (Zipursky, S. L., personal communication).

Development of the Mouse Cerebellum

As postnatal cerebellar development proceeds, granule cell neurons migrate from the external granular layer to the internal granular layer (Cowan, 1982). It has been postulated that the migrating neurons use the Bergmann glia, a subclass of astrocytes whose processes can span many cell layers, as guides in their movement (Rakic, 1982). Cell surface proteins such as L1 (Rathjen, 1984), and possibly N-CAM (Grumet, 1983) and Ng-CAM (Grumet, 1984) are thought to be involved in this interaction because antibodies directed against these molecules can disrupt

neuron-glia associations in vitro. The L1 molecule(s) is particularly intriguing because it is first expressed on post-mitotic
neurons just prior to their migration and because this temporal
specificity results in the formation of an increasing gradient of
L1 expression from the outer to the inner region of the external
granule cell layer (Schachner et al., 1983). The L1 antigen(s) is
a glycoprotein which runs on SDS-PAGE as two species: 200KD and
140KD (Schachner et al., 1983). In collaboration with the
Schachner group in Heidelberg, we have purified each of these
forms and obtained amino terminal sequence information from them.
These data have helped elucidate the structural relationships
among the polypeptides and will be invaluable in future homology
comparisons with the families of CAMs. In addition, these data
have been used to design oligodeoxynucleotide probes, which have
now been synthesized in the microchemical facility and are being
used to screen cDNA and genomic libraries.

DISCUSSION

    The conception and development of the microchemical facility
arose from a desire to understand developmental processes at the
genetic and molecular level. The instrumental tools to accomplish
this now exist and are being used not only in our own studies of
cell surface receptors, but in many other areas of biology. The
paradigm presented in Figure 1 illustrates ways in which one may
apply the current technology to biological systems. The loops of
the paradigm are continuous, illustrating the fact that one can

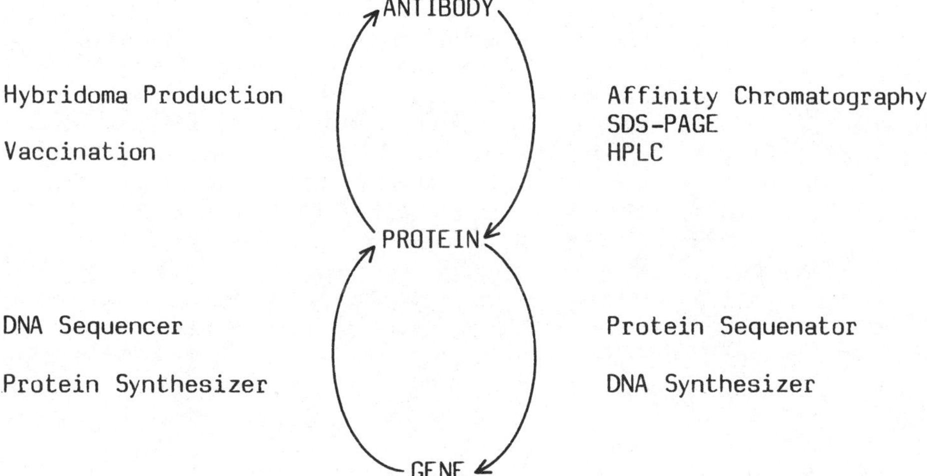

Fig. 1.   A paradigm for the application of microchemical instru-
          mentation to the study of proteins and genes.

initiate studies at any point and move to any other point one
desires. The instruments that have been developed facilitate this
process.

In the future we expect to extend our protein sequencing
sensitivity into the femtomole (nanogram) range by further
modifications in the Edman chemistry and by the incorporation of
the miniaturized mass spectrometer on line with the sequenator.
Special techniques are now being developed in the microchemical
facility to handle the minute quantities of protein this technology can utilize. These methods are designed to miniaturize and
simplify the preparation of sample proteins or peptides to effect
maximum yields.

Our study of the family of cell surface receptors (Mac-1,
LFA-1, etc.) which are differentially expressed on various blood
cell lineages provides a model approach which we plan to utilize
in the study of cell surface recognition during neural development. Our prediction is that the L1 protein of granule cell
neurons is one member of a family of molecules involved in
cellular recognition during neural development. We expect that
many other important molecules (and gene families) will be found
as a result of continuing efforts to understand cell recognition
processes, embryogenesis and development.

REFERENCES

Barany, G. and Merrifield, R. B. (1980). Solid-phase peptide
    synthesis, in: "The Peptides," Vol. II, E. Gross and J.
    Meienhofer, eds. Academic Press, New York.
Braunitzer, G. (1977). Pehr Victor Edman, in: "Solid Phase
    Methods in Protein Sequence analysis," A. Previero and
    M.-A. Coletti-Previero, eds. Elsevier/North-Holland
    Biomedical Press, Amsterdam.
Brown, J. P., Hewick, R. M., Hellström, I., Hellström, K. E.,
    Doolittle, R. F. and Dreyer, W. J. (1982). Human melanoma-
    associated antigen p97 is structurally and functionally
    related to transferrin. Nature 296:171-173.
Cowan, W. (1982). A synoptic view of the development of the
    vetebrate central nervous system, in: "Repair and
    Regeneration of the Nervous System," J. G. Nicholls, ed.
    Life Sciences Research Report 24. Springer Verlag, New York.
de Blas, A. L. (1984). Hybridoma technology applied to neuro-
    biological problems, in: "Current Methods in Cellular
    Neurobiology," J. L. Barker and J. F. McKelvy, eds.
    John Wiley and Sons, New York.
Dreyer, W. J. and Bennett, J. C. (1965). The molecular basis of
    antibody formation: A paradox. Proc. Natl. Acad. Sci. USA
    54:864-869.

Dreyer, W. J., Grey, W. and Hood, L. (1967). The genetic, molecular, and cellular basis of antibody formation: Some facts and a unifying hypothesis. Cold Spring Harbor Symp. Quant. Biol. **32**:353-367.

Dreyer, W. J., Kupperman, A., Boettger, H. G., Giffin, C. E., Norris, D. D., Gortch, S. L. and Theard, L. P. (1974). Automatic mass-spectrometric analysis: Preliminary report on development of a novel mass-spectrometric system for bio-medical applications. Clin. Chem. **20**:998-1002.

Dreyer, W. J. (1977). Peptide and protein sequencing method and apparatus. U.S. Patent #4,065,412.

Dreyer, W. J. (1984). Molecular evolution, antibody formation, and embryogenesis, in: "The Impact of Protein Chemistry on the Biomedical Sciences," A. N. Schechter, A. Dean, R. F. Goldberger, eds. Academic Press, New York.

Edman, P. and Begg, G. (1967). A protein sequenator. Eur. J. Biochem. **1**:80-91.

Fujita, S. C., Zipursky, S. L., Benzer, S., Ferrus, A. and Shotwell S. L. (1982). Monoclonal antibodies against the Drosophila nervous system. Proc. Natl. Acad. Sci. USA **79**:7929-7933.

Grumet, M., Rutishauser, U. and Edelman, G. M. (1983). Neuron-glia adhesion is inhibited by antibodies to neural determinants. Science **222**:60-62.

Grumet, M., Rutishauser, U. and Edelman, G. M. (1984). Two antigenically related neuronal cell adhesion molecules of different specificities mediate neuron-neuron and neuron-glia adhesion. Proc. Natl. Acad. Sci. USA **81**:267-271.

Hewick, R. M., Hunkapiller, M. W., Hood, L. E. and Dreyer, W. J. (1981). A gas-liquid solid phase peptide and protein sequenator. J. Biol. Chem. **256**:7990-7997.

Hood, L., Campbell, J. and Elgin, S. (1975). The organization, expression and evolution of antibody genes and other multi-gene families. Ann. Rev. Genet. 9:305-353.

Hood, L., Huang, H. V. and Dreyer, W. J. (1977). The area code hypothesis: The immune system provides clues to under-standing the genetic and molecular basis of cell recognition during development. J. Supramol. Struct. 7:531-559.

Huang, H., and Dreyer, W. J. (1978). Bursectomy in ovo blocks the generation of immunoglobulin diversity. J. Immunol. **121**:1738-1747.

Horvath, S. J., Firca, J., Graham, C., Hunkapiller, T., Caruthers, M., Hunkapiller, M. W., and Hood, L. (1984). An automated DNA synthesizer employing nucleoside 3' phosphoramidites. (Submitted for publication).

Hunkapiller, M. W., Lujon, E., Ostrander, F. and Hood, L. E. (1983). Isolation of microgram quantities of proteins from polyacrylamide gels for amino acid sequence analysis. Meth. Enzymol. **91**:227-236.

Katz, D. H. (1977). Lymphocyte Differentiation, Recognition, and Regulation. Academic Press, New York.

Kent, S. B. H. (1980). New aspects of solid-phase peptide synthesis, in: "Biomedical Polymers," E. P. Goldberg and A. Nakajima. eds. Academic Press, New York.

Kürzinger, K., Reynolds, T., Germain, R. N., Davignon, D., Martz, E., and Springer, T. A. (1981). A novel lymphocyte function-associated antigen (LFA-1): cellular distribution, quantitative expression, and structure. J. Immunol. 127:596-602.

Kürzinger, K. and Springer, T. A. (1982). Purification and structural characterization of LFA-1, a lymphocyte function-associated antigen, and Mac-1, a related macrophage differentiation antigen. J. Biol. Chem. 257:12412-12418.

Leder, P. (1982). The genetics of antibody diversity. Scientific Am. 246:102-115.

Maniatis, T., Hardison, R. C., Lacy, E., Lauer, J., O'Connell, C., Quon, D., Sim, G. K. and Efstratiadis, A. (1978). The isolation of structural genes from libraries of eukaryotic DNA. Cell 15:687-701.

Milstein, C. and Lennox, E. (1980). The use of monoclonal antibody techniques in the study of developing cell surfaces. Curr. Top. Dev. Biol. 14:1-32.

Potter, M., Dreyer, W. J., Kuff, E. L., and McIntire, K. R. (1964). Heritable variation in Bence Jones protein structure in an inbred strain of mice. J. Mol. Biol. 8:814-822.

Rakic, P. (1982). The role of neuronal-glial cell interaction during brain development, in: "Neuronal-glial Cell Interrelationships," T. A. Sears, ed. Life Sciences Research Report 20. Springer Verlag, New York.

Rathjen, F. G. and Schachner, M. (1984). Immunocytological and biochemical characterization of a new neuronal cell surface component (L1 antigen) which is involved in cell adhesion. EMBO J. 3:1-10.

Schachner, M., Faissner, A., Kruse, J., Lindner, J., Meier, D. H., Rathjen, F. G. and Wernecke, H. (1983). Cell type-specificity and developmental expression of neural cell surface components involved in cell interactions and of structurally related molecules. Cold Spring Harbor Symp. Quant. Biol. 48: in press.

Springer, T. A., Galfré, G., Secher, D. S., and Milstein, C. (1979). Mac-1: a macrophage differentiation antigen identified by monoclonal antibody. Eur. J. Immunol. 9:301-306.

Trowbridge, I. S., and Omary, M. B. (1981). Molecular complexity of leukocyte surface glycoproteins related to the macrophage differentiation antigen Mac-1. J. Exp. Med. 154:1517-1524.

Zipursky, S. L., Venkatesh, T. R., Teplow, D. B. and Benzer, S. (1984). Neuronal development in the Drosophila retina: Monoclonal antibodies as molecular probes. Cell 36:15-26.

THE INFLUENCE OF NEURONAL-GLIAL INTERACTIONS ON

GLIA-SPECIFIC GENE EXPRESSION IN EMBRYONIC RETINA

P. J. Linser* and A. A. Moscona**

*C. V. Whitney Laboratory, Univ. Florida
St. Augustine, FL 32086
**Cummings Life Sci. Ctr., Univ. Chicago
Chicago, IL

INTRODUCTION

The role of cell-cell interactions in morphogenesis and
differentiation of neural cells and tissues has been increasingly the focus
of considerable research activity (especially in recent years). One of the
most popular model systems for investigating these problems is the
embryonic neural retina. It is widely accepted that interactions between
heterotypic cells in the retina influence various developmental processes
(Moscona, 1974; Moscona & Linser, 1983). Our studies have addressed
specifically the role of cell interactions in regulating the expression of
developmentally programmed gene products. Since the avian retina is
avascular, biochemical and other studies on this tissue are not
complicated by the presence of non-neural cells. Also, since the avian
retina contains only one kind of neuroglia, Muller cells (Cajal, 1973), it is
possible to investigate the influence of retinal neurons on gene
expressions in the glial cell compartment. In this paper, we review some
of our observations concerning the role of neuron-glia interactions in
regulating the expression of specific gene products which are sequestered
in the glial compartment of the mature retina.

Muller cell markers: Glutamine Synthetase and Carbonic anhydrase-C

The enzymes glutamine synthetase (GS) and carbonic anhydrase-C
retina are present in very high concentration in the chicken neural
(NR)   representing 0.5% and 3.0% of the total tissue protein,
respectively (Moscona & Linser, 1983). By immunostaining with
monospecific, polyclonal, or monoclonal antibodies to these enzymes, we
have demonstrated that both GS and CA-C are localized in the Muller glia
cells of the mature retina (Linser & Moscona, 1979; 1981a; 1984). Fig. 1

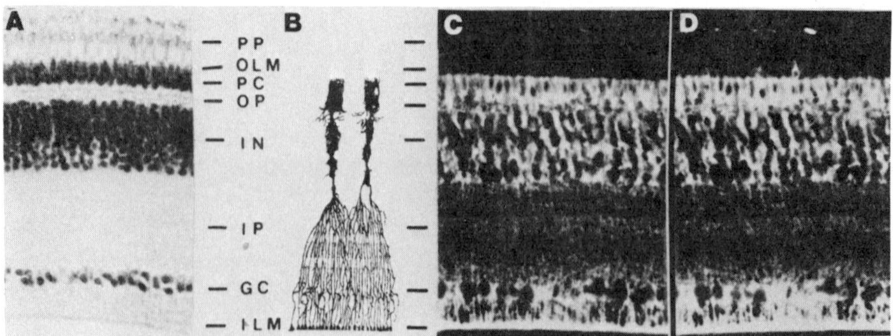

Fig. 1. Structure of mature chicken NR, Muller cell morphology, and simultaneous localization of CA-C and GS by double-label indirect immunofluorescence. (A) Hematoxylin-eosin-stained section of mature NR. PP, photoreceptor processes; OLM, outer limiting membrane; PC, photoreceptor cell layer; OP, outer plexiform layer; IN, inner nuclear layer; IP, inner plexiform layer; GC, ganglion cell layer; ILM, inner limiting membrane. (B) Morphology of mature avian Muller glia cells (after Cajal, 1973) showing numerous arborizations. (C and D) Same section of NR double-immunostained with antisera to CA-C (D) and GS (C). Immunofluorescence confined to Muller cells. The staining patterns of both enzymes closely overlap and correspond to the structural characteristics of Muller cells (x240). For detection of CA-C the section was treated with rabbit antiserum to CA-C, followed by rhodamine-conjugated goat anti-rabbit IgG Fab' fragments. For detection of GS, mouse antiserum to GS was used, followed by fluorescein isothiocyanate-conjugated rabbit anti-mouse IgG. (Modified from Linser and Moscona, 1981a).

shows the morphology of mature Muller cells of adult chicken retina and immunohistochemical staining for GS and CA-C, demonstrating that both enzymes are localized in these glia cells. Compartmentalization of both GS and CA-C in Muller cells is a general characteristic of most vertebrate retinas (Linser & Moscona, 1984). This undoubtedly reflects the important roles that these neuroglial cells play in retina development and function. GS is an integral constituent of the "small glutamate compartment" which is located in the glia cells and is directly involved in the metabolic turnover of amino acid neurotransmitters such as glutamate and GABA (Hamberger, et al., 1979; Stewart & Rosenberg, 1979). It has been postulated that Muller cells (in retina) and astroglial cells (in the CNS) take up these amino acids released during neuronal activity, convert them into glutamine through the action of GS and then release glutamine for uptake and reutilization by neurons (Moscona & Linser, 1983; Hamberger et al., 1979; Stewart & Rosenberg, 1979). This metabolic cycle functionally links neurons and glia so that the glia cells play an important role in neuronal activity. The detailed role of CA-C in retina function is less well understood, but this enzyme is of general

186

physiological importance in tissue and cell homeostasis in that it influences such parameters as water balance, pH, and ionic fluxes (Maren, 1967; 1976). CA-C may be significantly involved in early morphogenesis of the embryonic eye through its effects on intraoccular pressure (Linser & Moscona, 1981a; Maren, 1976). Hence, it appears that, whereas GS in Muller cells functions in a specific cooperative metabolic relationship between the glia and glutamatergic-GABAergic neurons, CA-C earmarks a more generalized role of the glia in the physiology of the retina and the eye.

Developmental expression of GS and CA-C

GS and CA-C show characteristic, but very different patterns of developmental expression during the ontogeny of the avian retina. Fig. 2 shows that the level of GS begins to increase very sharply only late in fetal development, during the period of functional maturation of the

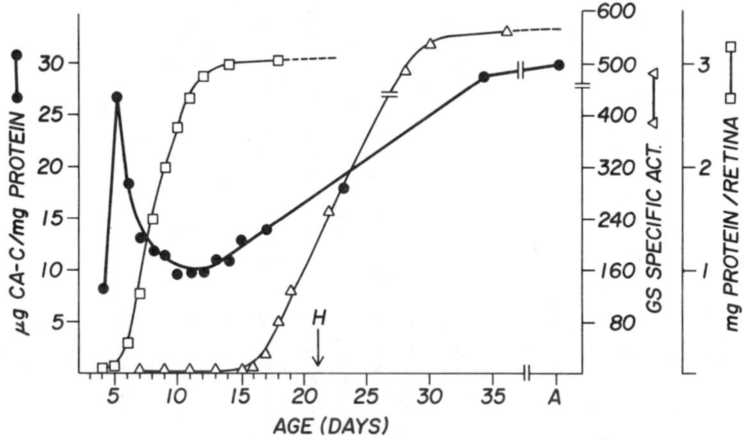

Fig. 2. Developmental profile of CA-C in the NR of chicken embryos compared with that of GS and with growth of the retina. ●-●, CA-C in retina, □-□, protein per retina; ▵-▵, GS specific activity; H, Hatching; A, adult stage. Units of GS-specific activity here were derived from an assay with improved sensitivity over that used in older work reviewed in this article. (From Linser & Moscona, 1981a).

retina (Piddington & Moscona, 1965; Moscona & Moscona, 1979). This
sudden rise of GS level represents one of the earliest documented
examples of corticosteroid-mediated induction of gene expression
(Moscona & Linser, 1983; Moscona & Piddington, 1966; Moscona, 1972). In
view of the above described role of GS in neuronal activity, it is not
surprising that this enzyme is regulated so that its rapid accumulation
begins concurrently with the onset of synaptic activity in the retina.
However, as in some other cases of hormonally elicited gene expressions
(Colowick and Kaplan, 1975), Muller cells are competent for GS induction
at a much earlier time in development than that at which it normally
occurs. Thus, if embryonic retina is prematurely exposed to the
appropriate corticosteroid, GS is precociously induced (Fig. 3; ref). It

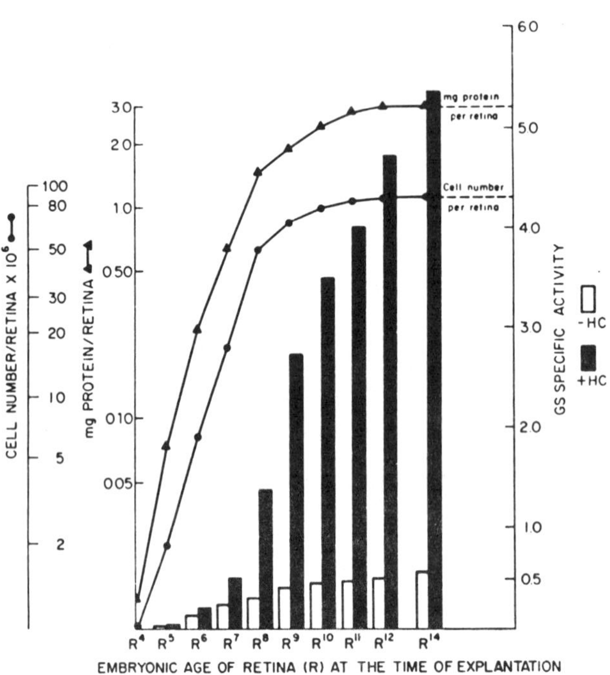

<u>Fig. 3.</u> Development of inducibility for GS in the NR of the chick embryo
between 5 and 14 days of incubation; relation to embryonic age, and
changes in cell number and in total protein per retina. Retinas dissected
from embryos were cultured for 24 hours in medium with cortisol (+HC) or
without it (-HC). The solid bars show the levels of GS activity induced in
retinas of different embryonic ages. The open bars show the levels of GS
in the absence of the steroid inducer. (From Moscona & Moscona, 1979).

188

should be stressed that both during normal development and in precocious induction (in vivo or in vitro), GS expression is always confined only to the Muller glial cells (Moscona & Linser, 1983a; Linser & Moscona, 1982)

The developmental pattern of CA-C expression in the retina is very different from that of GS (Fig. 2). In the retina of the chick embryo, CA-C is detectable already at a very early stage of development, and its concentration increases rapidly to a peak on about the 5th day of embryogenesis (Moscona & Linser, 1983; Linser & Moscona, 1981a). This rise of CA-C precedes by several days the appearance of competence for GS induction. Indeed, CA-C accumulation is "switched on" before there are any definitive specialized cell types in the retina, while it is still a simple neuroepithelium (Linser & Moscona, 1981a). During this early period, CA-C is immunohistochemically detectable in all of the retinoblasts. Later, the level of CA-C (relative to total protein) decreases somewhat for several days, but again begins to rise as the retina enters the post-mitotic period of functional cytodifferentiation (Moscona & Linser, 1983; Linser & Moscona, 1981a). During this period of definitive cell specialization, immunostaining for CA-C gradually disappears from the neurons, but intensifies in the Muller cells; eventually CA-C is found only in the glia cells. In the fundus (developmentally the most advanced region of the retina), CA-C is restricted to Muller cells, already by day 10-11 of embryonic development (Linser & Moscona, 1981a; 1982). We have recently found a similar progression of developmental changes with regard to the expression of both GS and CA-C, also in the mouse retina (Linser et al., 1984). In the mature retina of both species, GS and CA-C are markers for Muller cells[1], but during retina development the cellular localization and temporal programs of these top enzymes are dissimilar, which suggests that their expression is subject to different modes of regulation. One difference is that the developmental pattern of CA-C is not susceptible to modification by cortisol, while that of GS is. Another difference has to do with the effect of cell interactions. We will now briefly discuss the role of heterotypic cell interactions in regulating the expression of these two markers.

Cell-Cell Interactions

In the early work on cortisol-mediated induction of GS in chick embryo retina, it was noted that 3-dimensional structure of the tissue was essential for the expression of GS (Moscona, 1972; 1974). In those experiments, retinas from 8-16 day embryos were dissociated into single cells, by treatment with trypsin or mechanically, and the cells were

[1]It should be pointed out that in the mouse retina certain amacrine neurons continue to express CA-C in the mature tissue (Linser et al., 1984).

maintained either dispersed in monolayer cultures, or were compounded into aggregates in rotation cultures. Whole tissue fragments were cultured as controls (Moscona, 1972; 1974). In all cases, the culture medium contained cortisol as the inducer of GS. Fig. 4 shows the differences in GS inducibility in retina cells under these different conditions. The results showed that competance for GS induction depended on contact-interactions of the cells; in cell aggregates and in tissue fragments GS was induced to high levels, whereas monodispersed cells in monolayer culture showed almost no induction in spite of their continuous exposure to cortisol. Furthermore, it was also found that if embryonic retina cells were dissociated from retina tissue in which GS had been previously induced to a high level and were plated as a monolayer, they did not maintain the level of the enzyme (Moscona & Linser, 1983; Linser & Moscona, 1979).

The more recent findings that GS is localized in the retinoglia cells and that CA-C can serve as another marker for these cells enabled us to examine more closely the role of cell interactions in the expression of these enzymes. We approached this question by studying immunohistochemically GS and CA-C in embryonic Muller cells cultured under conditions which allowed control of cell-cell contact-relationships. The simplest approach was the use of adherent monolayer cultures of

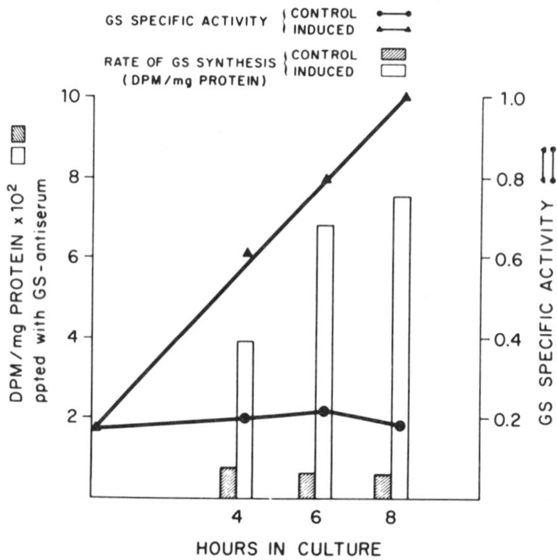

Fig. 4. Induction of GS by cortisol (+HC) in aggregates of NR cells, compared with absence of induction in cell monolayers. Also shown, GS induction in cultures of NR tissue. (From Moscona, 1974).

dispersed cells. By seeding such cultures with cells at different densities, the frequency of cell-cell contacts could be modulated from completely mono-dispersed cells, to the formation of multicellular aggregates that tended to detach from the substratum. In multicellular aggregates generated either in this way or in standard rotation cultures (Moscona & Linser, 1983; Linser & Moscona, 1979), GS is induced by cortisol in the Muller cells, and these cells also express CA-C (Moscona & Linser, 1983). Furthermore, if the aggregates were maintained up to a time equivalent to day 16 of embryonic age, CA-C expression became confined to the Muller cells, similar to development in vivo (Moscona & Linser, 1983).

In adherent monolayer cultures of post mitotic chick neural retina cells (from the fundus of eyes after the 10th day of development; Kahn, 1974), Muller cells assume a flattened epithelial morphology, whereas neurons are rounded with fine axon-like processes (Linser & Moscona, 1981b). Thus, in low density cultures, the Muller cell-derived gliocytes can be identified by their morphology. Immunocytochemical examination in such cultures showed that CA-C continued to be expressed in most of the gliocytic cells, whether cortisol was present or not. In contrast, GS was very rarely detected in these epithelial cells and only when they were closely juxtaposed with neurons (Linser & Moscona, 1983; Moscona & Linser, 1983). By seeding the cells at somewhat higher densities, it was possible to generate patches or sheets of epithelial gliocytes overlayed with clusters of neurons and neuronal cell processes (Fig. 5). Under such conditions, GS was inducible (with cortisol) and was immunocyto-chemically detected in numerous gliocytic cells, but always only if they were closely associted with neurons (Linser & Moscona, 1983; Moscona & Linser, 1983). In contrast, CA-C was found in epithelial cells irrespective of contact with neurons or the presence of cortisol. These results strongly suggested that the expression of the cortisol-elicited induction of GS in retinal glia cells requires neuron-glia contact-interactions, while CA-C expression in these cells persists independent of these same interactions, or the presence of cortisol.

To further test the hypothesis that neuron-glia contact interactions are involved in regulation of GS induction and expression in retinal glia cells, we prepared adherent monolayer cultures that consisted only of Muller cell-derived gliocytes and were devoid of neurons. This was achieved by plating dissociated 11-day embryonic retina cells at an intermediate cell density which was previously found to allow for maximal GS induction without generating spherical cell aggregates. The cultures were then treated with rabbit antiserum against purified retina cognin. The retinal cognin is a cell-surface glycosylated protein that is directly involved in the mechanism of retina cell-cell adhesion (Hausman & Moscona, 1976). In monolayer culture of post-mitotic embryonic retina cells, cognin is not expressed on the gliocytes (manuscript in

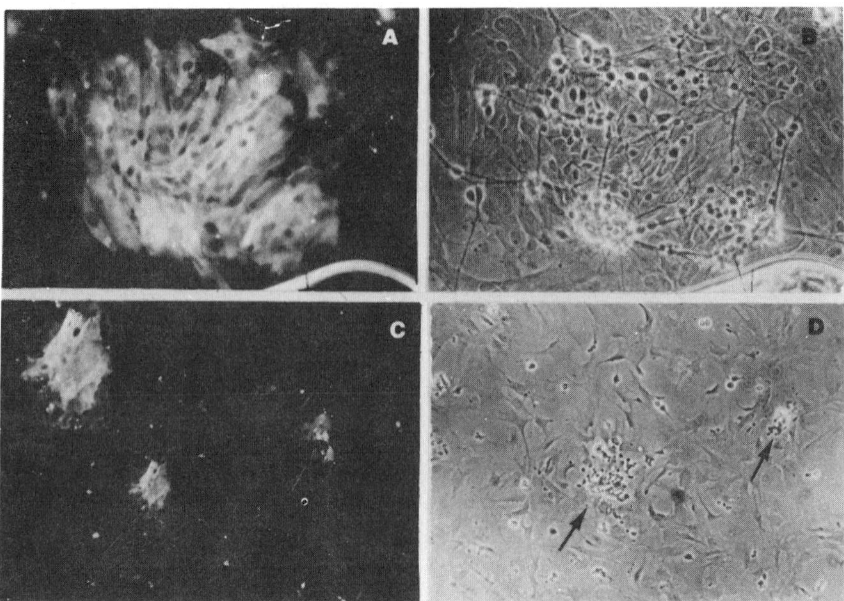

Fig. 5. Immunofluorescent analysis of GS localization in monolayer
cultures of nerual retina cells from 10-day (A, B) and 18-day (C, D)
embryos. Trypsin dissociated suspensions of single cells were plated at
1.3×10^6 cells/cm^2 in Medium 199 with 10% fetal bovine serum and
cortisol (0.33 ug/ml). Medium was changed daily; after 7 days, the
cultures were examined by indirect immunofluorescence for localization
of GS (A, C). Note staining for GS only in epithelioid cells (A, C, and
inset to C) and only when those cells are in contact with overlying
neurons, as evidenced by phase contrast micrographs of the same field (B,
D). (From Linser and Moscona, 1982).

preparation). After 2-3 days exposure to this antiserum (and endogenous
complement), virtually all the neuronal cells in the cultures were lytically
removed, leaving a monolayer of virtually pure gliocytes (Fig. 6; Linser &
Moscona, 1983). In other experiments, the neurons were removed by using
tetanus toxin and antiserum to the toxin (Linser & Moscona, 1983). When
the neuron-depleted monolayer cultures of glia-derived cells were treated
with cortisol and were examined immunocytochemically for GS and CA-C
expression, the cells immunostained intensely for CA-C but none
expressed GS (Fig. 6; Linser & Moscona, 1983). These results strongly
support the hypothesis that neuron-glia contact interactions are required
for the cortisol-mediated induction and expression of GS in retinal glia
cells, and that CA-C expression in these cells is regulated independently
of these requirements.

192

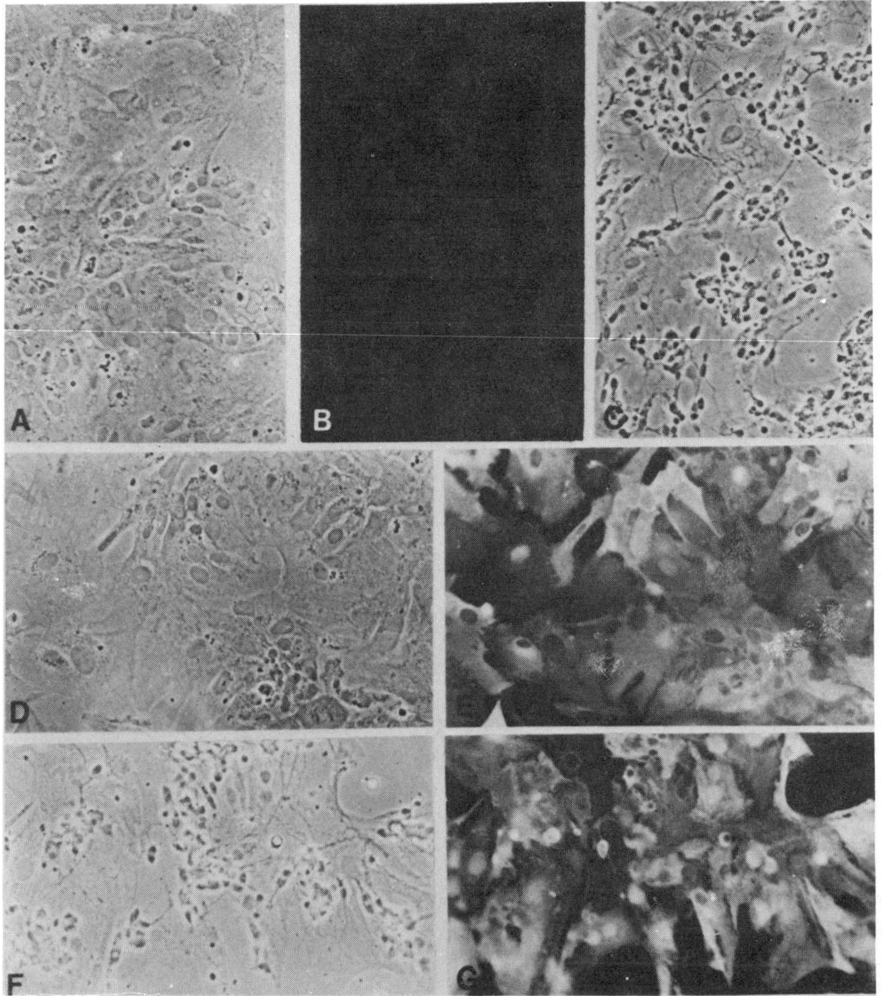

Fig. 6. Four-day cultures of retina cells (as in Fig. 1) in cortisol-
medium. (A) Culture depleted of neurons by treatment with
noninactivated antiserum to retina cognin (see text), showing only
gliocytes. (B) Same culture as in A, immunostained for GS; no detectable
enzyme. (C) Control culture for A, treated with peimmune serum; no
loss of neurons. (D) Neuron-depleted culture of gliocytes. (E) Same as
D, immunostained for CA-C, showing enzyme in virtually all the
gliocytes. (F) Regular culture containing neurons and gliocytes. (G)
Same as F, immunostained for CA-C x 300. (From Linser and Moscona,
1983).

The above observations raise questions as to the identity of the neurons involved in cell contact-regulation of GS inductions and expression in Muller cells, and of the nature of the cell-cell interactions exerting the regulatory effects. Glutamatergic and GABAergic neurons are likely candidates, since, as stated earlier, GS is specifically involved in the turnover of the amino acid neurotransmitters glutamate and GABA. Since our results have shown that, not only GS induction but also maintenance of high GS levels requires neuron-glia contact interactions, it should be possible to test which kinds of neurons are involved, by selective removal with specific neurotoxins. Of interest, in this context is the glutamate analogue kainic acid (KA). KA has been shown to be selectively toxic to neurons which are sensitive to excitation with glutamate (Olney et al., 1974). Furthermore, Schwarcz and Coyle have established by biochemical measurements that intraoccular injection of KA in chickens results in the selective loss of most cholinergic, GABAergic and glutamatergic neurons in the retina (Schwarcz and Coyle, 1977). KA selectively eliminated many of the interneurons of the inner nuclear layer, but apparently spared ganglion cells and photoreceptor cells (Schwarcz & Coyle, 1977). However, it was not reported whether Muller glial cells were affected. Therefore, we examined if KA treatment affected Muller cells morphologically, or in terms of GS and CA-C levels. Hatchling chicks received a single intraoccular injection of KA as described (Schwarcz & Coyle, 1977) in the left eye, while the right eye received vehicle injection as a control. The animals were sacrificed at various times thereafter, retinas were processed either for histology or for biochemical determination of the levels of GS, CA-C and choline acyltransferase (as a neuron marker). By histological criteria, our results were identical to those reported by Schwarcz and Coyle; 48 hrs after KA injection, there was marked neuropathy of the inner nuclear layer (INL) throughout the retina, and the inner plexiform layer (IPL) was dramatically reduced in size, whereas photoreceptor cell (PC) and ganglion cell (GC) layers were only minimally affected; from 48 hrs through 7 days, little further structural change was noted, except for further reduction in the width of the IN. Fig. 7 shows the appearance of control and KA-treated retinas at 7 days after treatment.

Immunohistochemical analysis of the two glial markers GS and CA-C showed that Muller cells were present; however, their morphological characteristics were drastically altered by exposure to KA. Muller glia cells are referred to as "Muller fibers" for obvious reasons (Fig. 7B,C); however, in KA-treated retinas, their fiber-like morphology was no longer apparent. In normal retina, the fine arborizations of Muller cells that run transversely through the layers from the inner limiting membrane (ILM) to the outer limiting membrane (OLM) are clearly visualized by immunostaining for GS or CA-C (Figs. 1, 7). In KA-treated retina, instead of these fine processes, there was more generalized staining throughout the retina layers (Fig. 7). The OLM and photoreceptor cell processes (PP)

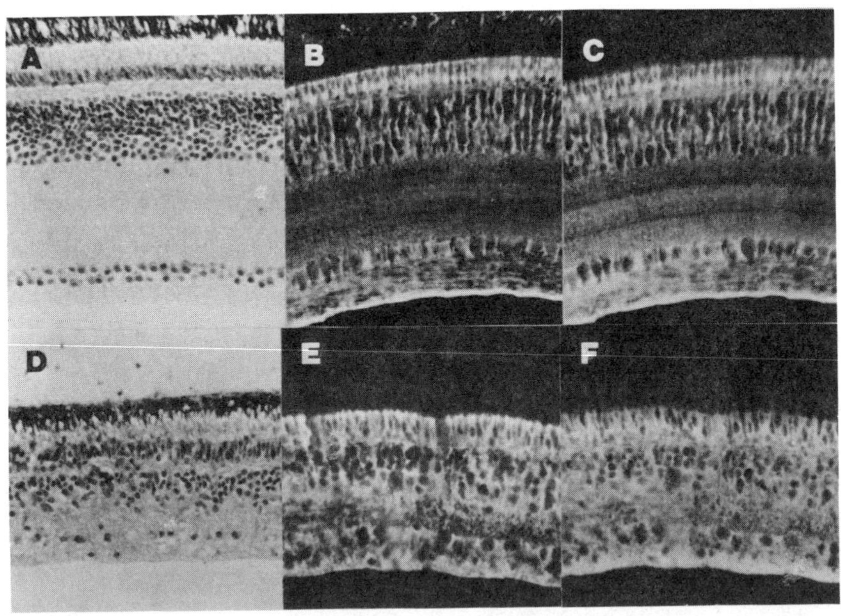

Fig. 7. Histological analysis of the effects of KA exposure on retina architecture. (A), hematoxylin and eosin stained section of normal hatchling chicken retina. (B & C), Section of normal retina stained for the simultaneous localization of GS (B) and CA-C (C) by double-label immunofluorescence (see legend Fig. 1). (D) Hematoxylin and eosin stained section of chicken retina obtained 7 days after administration of KA by intraoccular injection. Note reduction in cell number in the inner nuclear layer (for designation of retinal layers, see Figure 1). (E & F) Section of KA treated retina stained for GS (E) and CA-C (F) localization as in B & C.

were intact, the photoreceptor cells (PC), ganglion cells (GC), and remaining interneurons appeared in negative relief, as in the controls. The results indicated that in the KA-treated retina, Muller cells were retained, but their cytoplasmic extensions became diffusely re-oriented and now represented a major constituent of the contracted, neuron-depleted tissue (Fig. 7). Additionally, neurons remaining in the IN layer appeared to be primarily bipolar and/or horizontal, in that they were positioned on the OP side of the Muller cell perikaria. In the normal chicken retina, Muller cell perikaria form an approximate demarcation line between the amacrine neurons (at the IN side) and the bipolar and horizontal neurons (at the OP side) (Linser & Moscona, 1981c).

The biochemical effects of KA treatment in mature retina

corresponded to those detected morphologically; maximal changes
occurred within 48-72 hrs, with little or no further change thereafter.
The treatment reduced the wet weight and the total soluble protein of the
retina by approximately 20%. Fig. 8 shows the levels of CAT, CA-C and
GS 5 days after KA injection. As in the work of Schwarcz and Coyle, we
found neuronal CAT activity was reduced by more than 90%, indicating
the effectiveness of the treatment. On the other hand, glia-specific CA-
C concentration, expressed per unti total retina protein, was
approximately 25% higher than in untreated controls, due to the reduction
of total protein; the overall amount of CA-C per retina was equivalent to
that in control retinas. Thus, according to this glial marker, KA had no
destructive effect on the Muller cells. On the other hand, KA treatment
reduced the level of GS by 50% per unti protein (or roughly 75% per
retina). This reduction in GS level was noticeable already 48-72 hrs
following KA treatment and did not change later. Thus, although glia
cells were retained in the KA-treated retina, the expression of GS was
specifically and permanently "down-regulated" by the KA-caused
neuropathy.

While the results of the KA experiments are suggestive, it is not yet

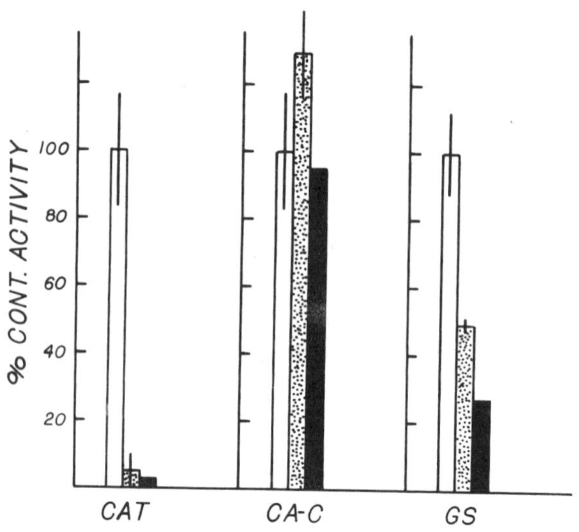

Fig. 8. Quantitation of choline acyltransferase (CAT) actively, GS
activity and CA-C concentration in control retinas (open bars) and in KA
treated retinas (stippled and closed bars). The stippled bars show the
activity (CAT and GS) or concentration (CA-C), per unit protein and the
closed bars show the same data expressed on a per retina basis. Error
bars indicate one standard deviation where N=7.

possible to decide exactly at what level the regulation of GS is affected.
It has been reported that KA does not prevent GS induction in the
embryonic retina, if KA is administered at an early stage before the
neurons have become susceptible to its toxic effects (Reif-Lehrer, 1981).
In fact, we found that KA had very little effect on GS inducibility in early
embryonic retinas. Accordingly, we suggest that in mature retina KA
reduces GS levels because of its toxicity to neurons that are involved in
cell contact-dependent regulation of GS expression in Muller cells. This
suggestion is consistent with the other evidence that GS expression is
regulated by neuron-glia contact interactions.

Finally, studies using the gliatoxic glutamate analogue, alpha
aminoadipic acid (AAA), suggest that still other glia cell functions
associated with the "small glutamate compartment" may be influenced by
neuronal input. AAA was shown to be specifically toxic to glia cells in
mature brain and retina (Olney et al., 1971; Pedersen & Karlsen, 1979).
Using organ cultures of retina tissue (chicken), we showed that AAA
destroys Muller cells also in late embryonic retina but not in the retina of
early embryos (Linser & Moscona, 1981b). Furthermore, we found that
sensitivity to this agent developed and increased concurrently with the
maturation of Muller glia cells, in parallel with the increase in inducibility
for GS (Fig. 9). The mechanism of AAA's selective toxicity for glia cells
is not clear, but the characteristic cytopathological effects of the drug
(extensive rapid edema and rarification of cytoplasm) lead Olney to
suggest that AAA acts upon some cell-surface receptor and thereby
causes an imbalance of ionic fluxes (Olney et al., 1971). Indeed, the small
glutamate compartment is associated with a high affinity glutamate
uptake mechanism which is coupled to ionic influx and efflux (Hamberger
et al., 1979; Stewart & Rosenberg, 1979). Thus, it may be that AAA
interacts with such a high affinity uptake mechanism and interferes in
some way with the coupled ionic fluxes, resulting in osmotic imbalance
and edema. These speculative considerations also raise the possibility
that the sensitivity of glia cells to AAA involves some type of cell-cell
interaction, possibly dependent on contact with neurons. In this respect,
perhaps the most suggestive observation made in examining Muller cell
sensitivity to AAA is the following: like responsiveness of Muller cells to
GS induction, their susceptibility to AAA also depends on their association
with neurons; if the retina is dissociated and the cells are maintained
dispersed in monolayer, the gliocytes become insensitive to AAA
cytolethality; cell reaggregation, i.e., restoration of 3-dimensional
contacts with neurons restores this sensitivity (unpublished observations).

Conclusions

The results of these studies indicate that disparate functions of
Muller cells involving the expression of CA-C and GS enzymes are
regulated independently, while there exists a functional relatedness

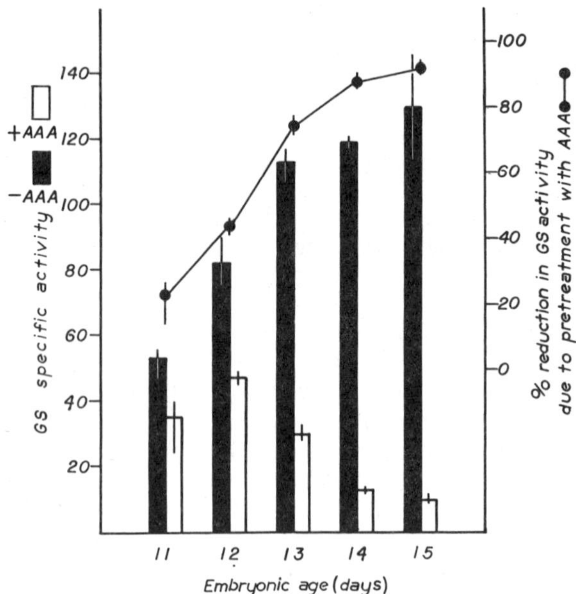

<u>Fig. 9</u>. Development of sensitivity to gliotoxic effects of AAA. Six day embryonic retinas were dissociated and reaggregated in suspension culture. At intervals, aggregates were exposed to AAA (200 ug/ml) for 48 hrs. The cell aggregates were transferred to cortisol-medium immediately following pretreatment with AAA and were assayed for GS 48 h later. Comparison is between the level of GS induced in aggregates pretreated with AAA (open bars) and in untreated aggregates (black bars). Points connected by line show the percentage reduction in GS inducibility due to pretreatment with AAA, as a function of embryonic age of the cells at assay time. All data represent means of 3 determinations. (From Linser and Moscona, 1981b).

between the regulation of GS inducibility and AAA sensitivity. It is very likely that ontogenetic cues which elicit specific differentiation changes in a given cell are temporally coordinated with the function of these changes: GS, which is required at high levels in Muller cells only late in development, when particular neurons become synaptically active is induced when such synaptic activity is beginning; CA-C is very likely required for the early stages of eye development and it appears already in the undifferentiated retinoblasts. The signals which trigger the accumulation of these two enzymes are also different: GS is induced by the combined effects of long-range cell-cell communiation via a hormonal messenger from the adrenal cortex (cortisol) and short-range cell-cell interactions involving heterotypic cell-cell contacts within the retina.

198

CA-C accumulation in the retina is switched on by as yet unknown cues
which differ both temporally and otherwise from those implicated in the
regulation of GS, and which may involve interactions among different
parts of the developing eye (Moscona & Linser, 1983; Linser & Moscona,
1981a; Linser & Moscona, 1982).

It has long been known that neuronal-glial interactions are
important for nervous tissue development and function. More recently,
there has been a growing number of examples suggesting the importance
of two-directional "information flow" between neurons and glia in the
differentiation of these cells (Holton & Weston, 1982; Aquayo et al., 1976;
Fisher, 1984). Neuron-glia interactions in the retina provide a useful
model for elucidating mechanisms of developmental cell-cell
communication. The present challenge is to identify and define the
molecular and genetic basis of such interactions and the detailed
mechanisms of their effects.

ACKNOWLEDGEMENTS

The original research described here is part of a research program
supported by Grant HD01253 from the National Institute of Child Health
and Human Development, and by Grant 1-733 from the March of Dimes-
Birth Defects Foundation.

REFERENCES

Aquayo, J. J., Epps, J., Charron, L. and Bray, G. M., 1976,
Multipotentiality of schwann cells in cross-anastomosed and grafted
myelinated and unmyelinated nerves, Brain Res., 104: 1-20.

Cajal, S. R., 1973, in "The Vertebrate Retina", R. W. Rodieck (ed)., W. H.
Freeman and Company, pp. 838-852.

Colowick, S. P. and N. O. Kaplan, 1975, Hormone Action, Part D: Isolated
Cells, Tissues and Organ Systems, in "Methods in Enzymology" Vol.
39, J. G. Hardman and B. W. O'Malley (eds.), Academic Press, New
York, 39.

Fisher, M., 1984, Neuronal-glial interactions and glial enzyme expression
in the mouse cerebellum, in "Gene Expression and Cell-Cell
Interactions in the Developing Nervous System", J. Lauder and P.
Nelson (eds.), Plenum Publishing Corp.

Hamberger, A. C., Han Chaing, G., Nylen, E. S., Scheff, S. W. and C. W.
Cotman, 1979, Glutamate as a CNS transmitter. I. Evaluation of
glucose and glutamine as precursors for the synthesis of
preferentially released glutamate, Brain Res. 168: 513-530.

Hausman, R. E. and A. A. Moscona, 1976, Isolation of retina-specific cell-
aggregating factor from membranes of embryonic neural retina
tissue, Proc Natl. Acad. Sci. USA 73: 3594-98.

Holton, B. and J. Weston, 1982, Analysis of glial cell differentiation in
peripheral nervous tissue. II. Neurons promote S-100 synthesis by
purified glial precursor cell populations. Develop. Biol. 89: 72-81.

Kahn, A. J., 1974, An autoradiographic analysis at the time of appearance
of neurons in the developing chick neural retina, Develop. Biol. 38:
30-40.

Linser, P. and A. A. Moscona, 1979, Induction of glutamine synthetase in
embryonic neural retina: localization in Muller fibers and
dependance on cell interactions, Proc. Natl. Acad. Sci. USA 76:
6476-6480.

Linser, P. and A. A. Moscona, 1981, Carbonic anhydrase C in the neural
retina: Transition from generalized to glia-specific cell localization
during embryonic development, Proc. Natl. Acad. Sci. USA 78: 7190-
94.

Linser, P. J. and A. A. Moscona, 1981, Induction of glutamine synthetase in embryonic neural retina: its suppression by the gliatoxic agent alpha-aminoadipic acid, Develop. Brain Res. 1: 103-119.

Linser, P. and A. A. Moscona, 1981, Developmental changes in the distribution of S-100 in avian neural retina, Dev. Neurosci. 4: 433-441.

Linser, P. J. and A. A. Moscona, 1982, Carbonic anhydrase-C in neural retina of the chick embryo: developmental changes in cellular localization, in "Embryonic Development, Part B: Cellular Aspects", M. Burger and R. Weber (eds.), Alan R. Liss Inc, NY, pp. 459-471.

Linser, P. and A. A. Moscona, 1983, Hormonal induction of glutamine synthetase in cultures of embryonic retina cells: requirement for neuron-glial contact interactions, Develop. Biol. 96: 529-534.

Linser, P. and A. A. Moscona, 1984, Variable CA-II compartmentalization in vertebrate retina, Ann. NY Acad. Sci.: (in press).

Linser, P. J., Sorrentino, M., and A. A. Moscona, 1984, Cellular compartmentalization of carbonic anhydrase-C and glutamine synthetase in developing and mature mouse neural retina, Develop. Brain Res.: (in press).

Maren, T., 1967, Carbonic anhydrase: chemistry, physiology and inhibition, Physiol Rev. 47: 595-781.

Maren, T. H., 1976, The rates of movement of Na^+, Cl^- and HCO_3^- from plasma to posterior chamber: effect of acetazolamide and relation to the treatment of glaucoma, Invest. Ophthal. 15: 356-364.

Moscona, A. A., 1972, Induction of glutamine synthetase in embryonic neural retina: a model for the regulation of specific gene expression in embryonic cells, F.E.B.S. Proc. 24: 1-23.

Moscona, A. A., 1974, Surface specification of embryonic cells: lectin receptors, cell recognition, and specific cell ligands, in "The Cell Surface in Development", A. A. Moscona (ed.), John Wiley & Sons, NY, pp. 67-99.

Moscona, A. A. and P. Linser, 1983, Developmental and experimental changes in retina glia cells: cell interactions and control of phenotype expression and stability, Curr. Top. Dev. Biol. 18: 155-188.

Moscona, M. and A. A. Moscona, 1979, Malformation of embryonic neural retina elicited by BrdU, <u>Differentiation</u> 13: 173-184.

Moscona, A. A. and R. Piddington, 1966, Stimulation by hydrocortisone of premature changes in the developmental pattern of glutamine synthetase in embryonic retina, <u>Biochim. Biophys. Acta</u> 121: 409-411.

Olney, J. W., Ho, O. L. and V. Rhee, 1971, Cytotoxic effects of acidic and sulphur containing amino acids on the infant mouse central nervous system, <u>Exp. Brain Res.</u> 14: 61-76.

Olney, J. W., Rhee, V. and O. L. Ho, 1974, Kainic acid: a powerful neurotoxic analogue of glutamate, <u>Brain Res.</u> 77: 507.

Pedersen, O. O. and R. L. Karlsen, 1979, Destruction of Muller cells in the adult rat by intravitreal injection of D, L- -aminoadipic acid. An electron microscopic study, <u>Exp. Eye Res.</u> 28: 569-575.

Piddington, R., and A. A. Moscona, 1965, Correspondence between glutamine synthetase activity and differentiation in the embryonic retina <u>in situ</u> and in culture, <u>J. Cell. Biol.</u> 27: 247-252.

Reif-Lehrer, L., 1981, Effects of kainic acid on glutamine synthetase activity in isolated chick embryo retina, <u>Supple Inves. Ophthalmol. Vis. Sci.</u> 20: 215.

Schwarcz, R. and J. T. Coyle, 1977, Kainic acid: neurotoxic effects after intraocular injection, <u>Invest. Ophthalmol. Vis. Sci.</u> 16: 141-148.

Stewart, R. M. and R. N. Rosenberg, 1979, Physiology of glia: glial-neuronal interactions, <u>Inter. Rev. Neuro.</u> 21: 275-309.

NEURON-GLIA INTERACTIONS AND GLIAL ENZYME EXPRESSION IN MOUSE

CEREBELLUM

Marilyn Fisher

Anatomy Department
University of Utah School of Medicine
Salt Lake City, Ut. 84132

INTRODUCTION

One of the major goals that developmental neurobiologists
share with those studying other developing systems is
understanding the role of cell interactions during morphogenesis
and cell differentiation. A variety of significant interactions
between neighboring cells occur during the formation of a normal
adult nervous system. In vertebrate embryos these interactions
begin with the induction of the neural epithelium by the
underlying notochord. Similar interactions are involved in the
morphogenesis and differentiation of auditory, lens and olfactory
placodes. Slightly later during brain development different types
of cell interactions come into play. For example, radial glia
serve as guides physically directing the migration of neurons, and
appropriate interactions (i.e. synapse formation) among certain
developing neuron populations are believed vital to the ultimate
survival of those neurons. Although these and other examples of
cellular interactions are widely accepted as important forces
during embryonic development, the mechanisms through which they
operate have proven to be elusive.

The developmentally regulated expression of the enzyme, _sn_-
glycerol-3-phosphate dehydrogenase (GPDH; EC 1.1.1.8) in the mouse
cerebellum appears to be a specific example where one cell type
exerts an influence over the metabolic properties of another cell
type. In this paper I shall review the experimental and
inferential bases supporting this notion and the features of GPDH
expression that recommend it particularly as a model for studying
many questions related to the control of gene expression during
development.

GPDH EXPRESSION IN NORMAL CEREBELLAR DEVELOPMENT

The enzyme, GPDH, is present in the form of an embryonic
isozyme with low levels of activity in many tissues of the embryo
and fetus. At various times during pre- and postnatal development
(depending upon the tissue), a genetically distinct adult isozyme
overshadows the embryonic isozyme (Kozak and Jensen, 1974; Kozak
et al., 1982). This event occurs in the mouse cerebellum
primarily during the second and third weeks after birth; a time
when extensive cell migration and differentiation take place
within the cerebellum. The shift in predominant isozyme types is
marked by a sixty-fold increase in enzyme specific activity that
directly follows increases in levels of translatable,
GPDH-specific mRNA and enzyme synthesis (Kozak and Ratner, 1980).
Immunocytochemical staining, using an antiserum specific for the
adult isozyme, revealed that the dramatic rise in enzyme activity
corresponds to the appearance of immunoreactive enzyme
concentrated in two cell populations (Fisher et al., 1981). High
levels of immunoreactive GPDH are seen in both Bergmann glia
(Golgi epithelial cells) and oligodendrocytes after postnatal day
12 (Figure 1). In addition to the intense staining of the cell
bodies, Bergmann glia impart homogeneous staining to the molecular
layer by virtue of their extensive veil-like processes that
surround the Purkinje cell dendrites. The homogeneous background
of staining is interrupted only by the scattered, unstained bodies
of stellate, basket and other cell types residing in that layer.
The radial Bergmann fibers appear as distinct, intensely stained
fibers when they fall within the plane of section. Staining of
the Bergmann glia and molecular layer is uniform throughout the
cerebellar cortex in the normal mouse.

GPDH EXPRESSION IN MUTANTS AND CHIMERAS

Having identified specific cell types as apparently
responsible for the observed dramatic shift in enzyme expression,
a more focused approach to identifying the factors that control
this pattern of enzyme expression is possible. Examination of a
large number of neurological mouse mutants identified several
having significantly reduced levels of GPDH activity. The mutants
identified by the preliminary screening carried out at The Jackson
Laboratory in collaboration with Dr. Les Kozak were: Lurcher;
Purkinje cell degeneration; reeler; staggerer and weaver.
Immunocytochemical studies revealed a corresponding loss of
immunoreactive GPDH, specifically from the molecular layer and
Bergmann glia cell bodies. Staining of oligodendrocytes in these
mutants appeared unaffected (Figures 2-4). In contrast, a mutant
with an oligodendrocyte deficiency, jimpy, showed normal staining
of the Bergmann glia and molecular layer but no immunoreactive
oligodendrocytes. Biochemically determined enzyme activity in

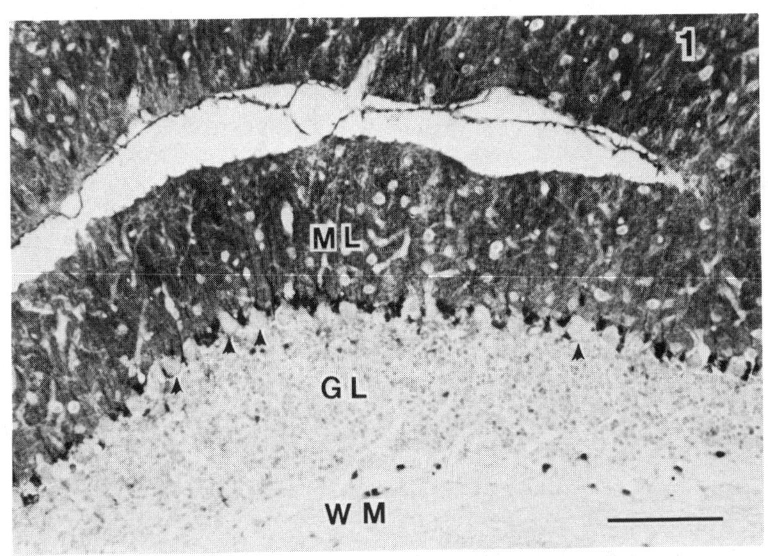

Figure 1. Cerebellar section from a normal 3 week old mouse immunostained for GPDH. The molecular layer (ML) is stained homogeneously except for cell bodies and Purkinje cell dendrites which appear as clear areas. Bergmann glia somas are visible as intensely stained spots interspersed among the unstained Purkinje cell somas (arrowheads). Intensely stained radial Bergmann fibers can be seen crossing the molecular layer. A few darkly stained oligodendrocytes are visible within the granule cell layer (GL) and white matter (WM). Calibration bar = 100 μm.

jimpy mice was only slightly below that of normal littermates underscoring the relatively minor contribution of this cell type to total cerebellar enzyme levels (Fisher et al., 1981).

There is little evidence to suggest that the loss of GPDH from the Bergmann glia in these mutants results in each case from an intrinsic defect in the glial cell population. Defects in the Bergmann glia in weaver mice were reported by Rakic and Sidman (1973) who found reduced numbers and morphological abnormalities of these cells. However, it appears from the studies of Sotelo and Changeux (1974) and Bignami and Dahl (1974) that those defects may be secondary and may be related to the genetic background upon which the mutation was operating. Thus, although no consistent defect of the Bergmann glia has been described in these mutants,

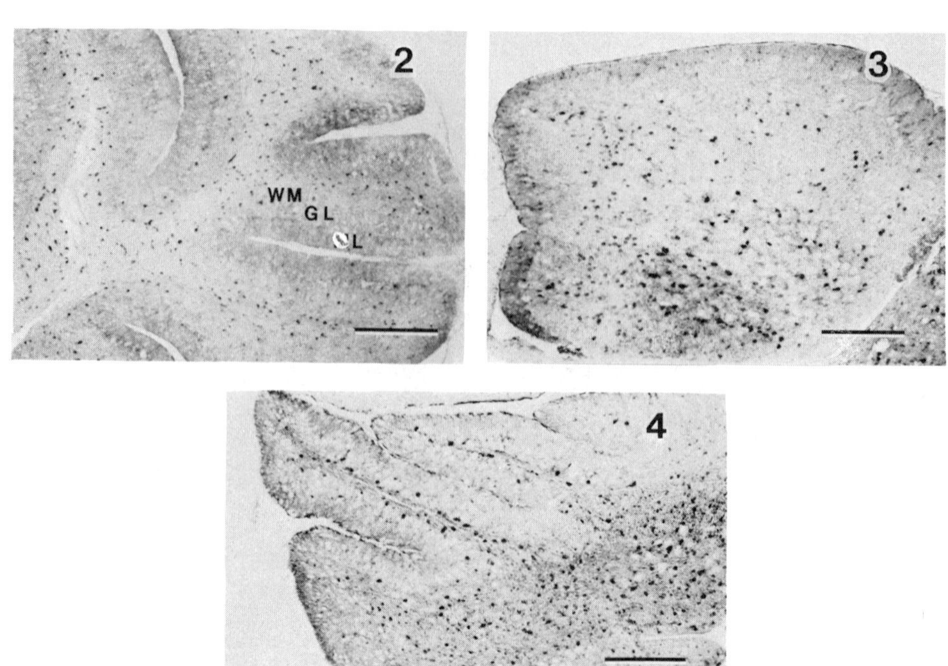

Figure 2. Cerebellar section from a 4 week old Lurcher mutant
immunostained for GPDH. The molecular layer (ML) is not
intensely stained as it was in the normal. Purkinje cells
are absent from the boundary between the molecular and
granule cell layers. Note also, at this boundary, the
absence of the row of intensely stained Bergmann glia
somas. The intensely stained cells that are present are
oligodendrocytes scattered throughout the granule cell
layer (GL) and white matter (WM). Calibration bar = 200
μm.

Figure 3. Cerebellar section from a 3 week old reeler mutant
immunostaineed for GPDH. The thin molecular layer around
this unfoliated cerebellum is only lightly stained
compared with the normal. Some staining is apparent in
the anterior portion of the molecular layer (left side of
photo). While a few Purkinje cells appear in their normal
location, the majority lie deep to the rudimentary granule
cell layer. Again, there are few if any intensely stained
Bergmann glia somas visible along the boundary between
molecular and granule cell layers. Intensely stained
oligodendrocytes are visible scattered throughout the
deeper layers of this mutant cerebellum. Calibration bar
= 200 μm.

Figure 4. Cerebellar section from a 3 week old weaver mutant

(continued)

---

the possibility of intrinsic glial defects has not been ruled out.

In contrast to the lack of obvious glial defects in the
mutants found to have reduced GPDH levels all have serious
Purkinje cell defects and most have at least some reduction in
granule cell numbers. It is possible, then, that the effects of
these mutations on expression of GPDH in Bergmann glia may be
secondary to their effects on neuronal cells, particularly
Purkinje cells, since this cell type is consistently severely
affected in each of the mutants. The mutants Lurcher and
Purkinje cell degeneration lose their entire Purkinje cell
populations during the first month (Caddy and Biscoe, 1979; Mullen
et al., 1976). The Purkinje cells in reeler are severely
malpositioned, and in the mutants reeler, staggerer and weaver,
their dendritic trees are severely stunted (Sidman, 1968; Bradley
and Berry, 1978).

Why should expression of GPDH in Bergmann glia be influenced
by Purkinje cells? The enzyme, GPDH, is ubiquitous and functions
in the cellular synthesis of phospholipids and in energy
metabolism through the glycerol phosphate cycle (Lin, 1977). In
the cerebellum two cell populations have high concentrations of
immunoreactive GPDH: Bergmann glia and oligodendrocytes. Both of
these cell types might be expected to engage in an especially high
level of phospholipid synthesis as a consequence of their
extensive plasma membrane production and maintenance.
Oligodendrocytes elaborate myelin sheaths whereas Bergmann glia
ensheath the vast Purkinje cell dendritic networks (Palay and
Chan-Palay, 1974). High levels of GPDH in oligodendrocytes have
been linked to the process of myelination (de Vellis et al.,
1977). Furthermore, at least in the peripheral nervous system,
myelin production directly results from an interaction between
Schwann cells and the axons they contact (Aguayo et al., 1976a,b).
A similar interaction between the Bergmann glia and Purkinje cells
may be active during ensheathment of the Purkinje cell dendrites.
In that case, one would predict that any circumstance causing
substantial loss of Purkinje cell dendritic volume would be
reflected in reduced GPDH activity, specifically within the
Bergmann glia. Detailed study of three mutations with substantial
Purkinje cell loss (including one that was not identified in the

preliminary screening) supports this prediction.

In the mutant Lurcher, where Purkinje cell loss begins very early, GPDH activity is reduced to 60% of that of normal littermates at postnatal day 18 (the earliest time measured). The relative difference in enzyme activity between mutant and controls increases until by 6 weeks postnatal the mutants have only 20% of the normal level of GPDH activity. In the mutant Purkinje cell degeneration, where the onset of Purkinje cell loss is slightly delayed compared to Lurcher, loss of GPDH activity is also delayed. There is a rapid loss of immunoreactive GPDH from the Bergmann glia cell bodies by 6 weeks of age, but enzyme only gradually disappears from the processes within the molecular layer. The difference in the details of enzyme loss following Purkinje cell death in these two mutants may be due to differences in the genetic backgrounds on which these mutants are carried and probably indicate that additional factors besides the presence of Purkinje cells can influence expression of GPDH in the Bergmann glia.

The most interesting mutant studied is nervous. This mutant differs from Lurcher and Purkinje cell degeneration in that most, but not all, of the Purkinje cell population dies. In these mice 50% or more of the Purkinje cells in the vermis and up to 90% in the lateral hemispheres die by 2 months of age (Sidman and Green, 1970). The resulting cerebellum has patches of cortex that are sparsely populated with Purkinje cells and patches that have none. As Figure 5 shows, some patches of molecular layer are intensely immunoreactive for GPDH and others are apparently lacking in GPDH. Close examination of the patches shows that intense GPDH immunoreactivity exists only where Purkinje cells are present (Figure 6). The possibility that Bergmann glia, like the Purkinje cells, may die in this mutant has been ruled out by the demonstration that Bergmann glia are present throughout the cortex; this evidence was obtained by immunostaining with an antiserum specific for the glial fibrillary acidic protein (Fisher, 1984).

Similarly, patchy staining patterns also occurred in chimeric mice made in collaboration with Dr. Richard Mullen using either of the mutants, Lurcher or Purkinje cell degeneration. Purkinje cell death in both mutants results from the intrinsic action of those genes within the Purkinje cells (Wetts and Herrup, 1982; Mullen, 1977). Thus, chimeric mice made by aggregating a normal embryo with an embryo of either mutant genotype will have patches where genetically normal Purkinje cells are present and Purkinje cell-free patches where genetically mutant cells have died. As in the nervous mutant, there is always a positive correlation in these chimeras between the presence of Purkinje cells and intense GPDH immunoreactivity in the Bergmann glia and molecular layer

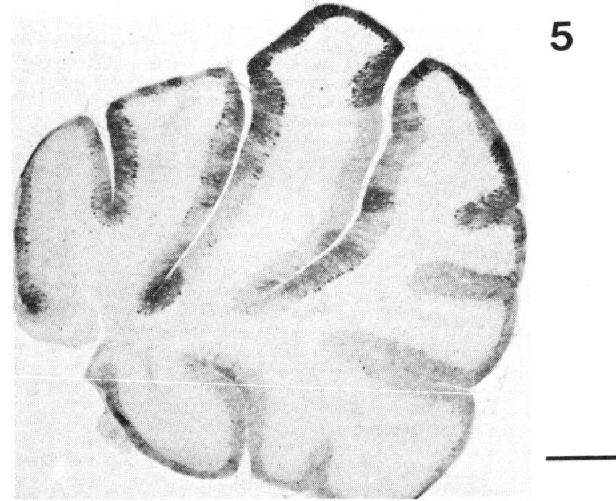

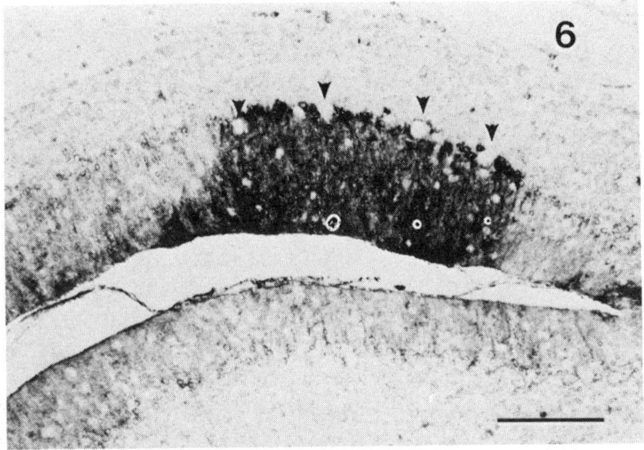

Figure 5. Cerebellar section from a 7 month old nervous mutant
immunostained for GPDH. The molecular layer has regions
of intense staining as well as regions of very light
staining. Calibration bar = 300 μm.

Figure 6. Region from a section semiadjacent to that illustrated
in Figure 5. Unstained Purkinje cell somas (arrowheads)
are present in the intensely stained patch of molecular
layer but are absent from the unstained patches to either
side of and opposite the stained patch. Calibration bar =
100 μm.

(Figure 7). Again, the patchy GPDH staining seen in the chimeras
contrasts with the uniform distribution of Bergmann fibers
demonstrated by immunostaining for the presence of glial
filaments.

These results suggest two important features of the
interaction between the Bergmann glia and Purkinje cells. First,
the abrupt boundaries between GPDH-stained and unstained regions
of molecular layer seen in nervous and chimeric cerebella (see
Figures 5 and 7) argue against the involvement of a diffusible
factor and in favor of direct cell contacts. Second, the
observation that the influence of an isolated Purkinje cell is
restricted to an area similar in size to the extent of its own
dendritic tree supports the notion that the Purkinje cell
dendrites themselves are involved.

GPDH EXPRESSION AS A MODEL OF GENE CONTROL

As mentioned above, very little is known about the mechanisms
through which interacting cells influence one another during
development. There are other examples of interacting systems in
the developing nervous system, such as axon "induction" of
myelination (Aguayo et al., 1976a,b) and the neuron-glia
interaction involved in glutamine synthetase expression in the
retina (Linser and Moscona, 1979, and this volume). These are
both important and interesting systems that are being actively
studied. They have been successfully exploited mainly because
they are amenable to study in tissue culture. Glial GPDH
expression offers a rather unique combination of additional
features that will permit the study of many questions related to
gene control using a variety of techniques and approaches.

GPDH expression in reaggregating cell cultures from young
postnatal mouse cerebella exhibits a pattern similar to that seen
in vivo (Kozak, 1977). Thus, this phenomenon is likewise
accessible for _in vitro_ study. It should be possible in culture
to design appropriate experiments to determine whether the
interaction between neurons and glia is mediated through direct
cell contact and, hopefully, to identify relevant cell surface
bound or associated components.

A second useful feature is the availability of much
information concerning the genetics of GPDH expresssion. There
are strain differences in cerebellar GPDH expression that involve
two loci, one of which is unlinked to Gdc-1, the structural locus
of GPDH (Kozak, 1972). Furthermore, there are biochemically
distinct allelic variants of both the adult and embryonic isozymes
(Kozak, 1972; Kozak and Erdelsky, 1975; Chan and Thompson, 1981;
Kozak et al., 1982). The availability of these genetic tools will
permit the exploration of additional questions concerning finer
levels of regulation.

Finally, a cDNA clone of the structural locus for GPDH is now
available (Kozak and Birkenmeier, 1983). This advance will enable
studies to determine the precise molecular level at which enzyme
expression is being regulated in this system.

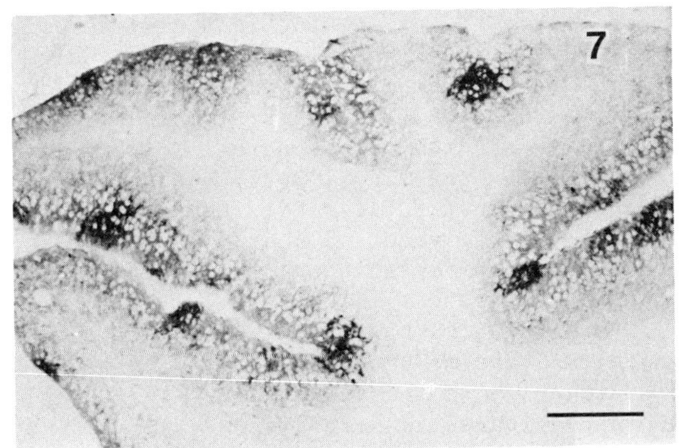

Figure 7. Cerebellar section from a 9 month old Purkinje cell
    degeneration chimera immunostained for GPDH. Patches of
    intense immunoreactivity stand out in the otherwise
    unstained molecular layer. Calibration bar = 200 μm.

Thus, in the case of GPDH expression, there seems to be a
rather unique combination of tools available. Hopefully, with
these tools we can gain an understanding at the molecular level of
at least one mechanism by which interactions between cells can
influence gene expression.

ACKNOWLEDGMENTS

Various aspects of the work summarized here was supported by
the following NIH grants: NS17790 (to M. F.); NS16156 (to Dr. R.
J. Mullen); HD06712 (to Dr. L. P. Kozak) and Training Grant
GM07386 (to The Jackson Laboratory).

REFERENCES

Aguayo, A. J., Charron, L., and Bray, G. M., 1976a, Potential of
    Schwann cells from unmyelinated nerves to produce myelin:
    A quantitative ultrastructural and radiographic study, J.
    Neurocytol., 5:565.
Aguayo, A. J., Epps, J., Charron, L., and Bray, G. M., 1976b,
    Multipotentiality of Schwann cells in cross-anastomosed
    and grafted unmyelinated nerves. Quantitative microscopy
    and radioautography, Brain Res., 104:1.
Bignami, A., and Dahl, D., 1974, The development of Bergmann glia
    in mutant mice with cerebellar malformation: reeler,
    staggerer and weaver. Immunofluorescence study with
    antibodies to the glial fibrillary acidic protein, J.
    Comp. Neurol., 155:219.

Bradley, P., and Berry, M., 1978, Purkinje cell dendritic tree in
     mutant mouse cerebellum. A quantitative Golgi study of
     weaver and staggerer mice, <u>Brain Res</u>., 142:135.
Caddy, K. W. T., and Biscoe, T. J., 1979, Structural and
     quantitative studies on the normal C3H and Lurcher mutant
     mouse, <u>Philos</u>. <u>Trans</u>. <u>Roy</u>. <u>Soc</u>. <u>London</u> , 287:167.
Chan, A. K., and Thompson, E. A., 1981, Appearance of a second
     form of hepatic glycerol-3-phosphate dehydrogenase during
     neonatal development in the mouse, <u>Arch</u>. <u>Biochem</u>.
     <u>Biophys</u>., 207:96.
de Vellis, J., McGinnis, J. F., Breen, G. A. M., Leveille, P.,
     Bennett, K., and McCarthy, K., 1977, Hormonal Effects on
     differentiation in neural cultures, <u>in</u>: "Cell Tissue
     and Organ cultures in Neurobiology," S. Federoff and L.
     Hertz, eds., Academic Press, New York.
Fisher, M., 1984, Neuronal influence on glial enzyme expression:
     Evidence from mutant mouse cerebella, Submitted.
Fisher, M., Gapp, D. A., and Kozak, L. P., 1981,
     Immunohistochemical localization of <u>sn</u>-glycerol-3-
     phosphate dehydrogenase in Bergmann glia and
     oligodendroglia in the mouse cerebellum, <u>Dev</u>. <u>Brain Res</u>.,
     1:341.
Kozak, L. P., 1972, Genetic control of a-glycerol phosphate
     dehydrogenase in mouse brain, <u>Proc</u>. <u>Natl</u>. <u>Acad</u>. <u>Sci</u>.
     <u>USA</u>, 69:3170.
Kozak, L. P., 1977, The transition from embryonic to adult isozyme
     expression in reaggregating cell cultures of mouse brain,
     <u>Dev</u>. <u>Biol</u>., 55:160.
Kozak, L. P., and Birkenmeier, E. H., 1983, Mouse
     <u>sn</u>-glycerol-3-phosphate dehydrogenase: Molecular cloning
     and genetic mapping of a cDNA sequence, <u>Proc</u>. <u>Natl</u>.
     <u>Acad</u>. <u>Sci</u>. <u>USA</u>, 80:3020.
Kozak, L. P., Burkart, D. L., and Hjorth, J. P., 1982, Unlinked
     structural genes for the developmentally regulated
     isozymes of <u>sn</u>-glycerol-3-phosphate dehydrogenase in
     mice, <u>Dev</u>. <u>Genet</u>., 3:1.
Kozak, L. P., and Erdelsky, K. J., 1975, The genetics and
     developmental regulation of L-glycerol-3-phosphate
     dehydrogenase, <u>J</u>. <u>Cell</u>. <u>Physiol</u>., 85:437.
Kozak, L. P., and Jensen, J. T., 1974, Genetic and developmental
     control of multiple forms of L-glycerol-3-phosphate
     dehydrogenase, <u>J</u>. <u>Biol</u>. <u>Chem</u>., 249:7775.
Kozak, L. P., and Ratner, P. L., 1980, Genetic regulation of
     translatable mRNA levels of mouse <u>sn</u>-glycerol-3-
     phosphate dehydrogenase during development of the
     cerebellum, <u>J</u>. <u>Biol</u>. <u>Chem</u>., 255:7589.
Lin, E. C. C., 1977, Glycerol utilization and its regulation in
     mammals, <u>Ann</u>. <u>Rev</u>. <u>Biochem</u>., 46:765.
Linser, P., and Moscona, A. A., 1979, Induction of glutamine
     synthetase in embryonic neural retina: Localization in

Muller fibers and dependence on cell interactions, <u>Proc.
Natl. Acad. Sci. USA</u>, 76:6476.
Mullen, R. J., 1977, Site of pcd gene action and Purkinje cell
mosaicism in cerebella of chimaeric mice, <u>Nature</u>,
270:245.
Mullen, R. J., Eicher, E. M., and Sidman, R. L., 1976, Purkinje
cell degeneration, a new neurological mutation in the
mouse, <u>Proc. Natl. Acad. Sci. USA</u>, 73:208.
Palay, S. L., and Chan-Palay, V., 1974, "Cerebellar Cortex,"
Springer-Verlag, New York.
Rakic, P., and Sidman, R. L., 1973, Weaver mutant mouse
cerebellum: Defective neuronal migration secondary to
abnormality of Bergmann glia, <u>Proc. Natl. Acad. Sci</u>
<u>USA</u>, 70:240.
Sidman, R. L., 1968, Development of interneuronal connections in
brains of mutant mice, <u>in</u>: "Physiological and
Biochemical Aspects of Nervous Integration," F. D.
Carlson, ed., Prentice-Hall Inc., Englewood Cliffs, N. J.
Sidman, R. L., and Green, M. C., 1970, "Nervous", a new mutation
with cerebellar disease, <u>in</u>: "Les Mutants Pathologiques
chez L´animal," M. Sabourdy, ed., Centre National de La
Recherche Scientifique, Paris.
Sotelo, C., and Changeux, J. P., 1974, Bergmann fibers and
granular cell migration in the cerebellum of homozygous
weaver mutant mouse, <u>Brain Res</u>., 77:484.
Wetts, R., and Herrup, K., 1982, Interaction of granule, Purkinje
and inferior olivary neurons in Lurcher chimaeric mice.
I. Qualitative studies, <u>J. Embryol. Exp. Morph</u>.,
68:87.

DEVELOPMENT OF GABA-ERGIC SYSTEM IN RAT VISUAL CORTEX

J.R. Wolff, V.J. Balcar, T. Zetzsche, H. Böttcher,
D.E. Schmechel* and B.M. Chronwall**

Dept. of Anatomy, Univ. of Göttingen, D3400, F.R.G.
and *Dept. of Neurol., Duke Univ., Durham, U.S.A.
**present address: Exper. Therap. Branch, NIH
Bethesda, MD, U.S.A.

INTRODUCTION

In 1956 Eugene Roberts stated, "Perhaps the most difficult
question to answer would be whether the presence in the grey
matter of the central nervous system of uniquely high
concentrations of γ-aminobutyric acid and the enzyme which forms
it from glutamic acid has a direct or indirect connection to
conduction of the nerve impulse in this tissue". Thirty years
later γ-aminobutyrate (GABA, 4-aminobutyrate) is universally
accepted as the major inhibitory synaptic transmitter. This review
is trying to explore its additional roles, beyond a mediation of
synaptic transmission. Specifically, we are examining a
possibility that GABA, on account of its characteristic chemical
properties in interaction with excitable membranes, mediates
communication among differentiating neurons, thus influencing
expression of neuronal functions, making the orderly development
of synaptic connections possible.

It has been stated that during the ontogenetic development
the neurons forming the local circuits originate and differentiate
later than corresponding projection neurons (Jacobson, 1978).
This is not in accord with the observations that in the visual
cortex of rats, GABA-accumulating neurons, which are typical
local-circuit neurons in this tissue (Chronwall and Wolff, 1978a),

---

Acknowledgement: Supported by Deutsche Forschungsgemeinschaft
Grants No. SFB 33 -E5, -C15, Wo 279/5-1

start appearing already on embryonic day 16 (E16; Chronwall and
Wolff, 1980) and GABA level at birth is about half that of the
adult cortex (Coyle, 1982). This apparent discrepancy indicated
that GABA may have a specific role during the development and an
hypothesis to this effect was formulated (Wolff, 1981).

We have studied ontogenetic development of GABA-operated
system in rat visual cortex, an anatomically well defined part
of the cortex where most of the synaptic inhibition is mediated
by GABA and all inhibitory GABA-ergic (GABA-releasing) synapses
originate from neurons intrinsic to the region. By confining our
studies to a relatively small part of the cerebral cortex we also
hoped to minimize effects of temporo-spatial gradients in the
cortical differentiation (His, 1904; Raedler and Sievers, 1975)
and to obtain as clear a picture as possible of the development
of various components of the GABA-ergic inhibitory system.

GABA-ergic neurons have been visualized by two methods:
by autoradiography with ^{3}H-GABA (Hökfelt and Ljungdahl, 1972)
and by immunocytochemistry using antibodies against glutamate
decarboxylase (GAD), a GABA-synthesizing enzyme (Ribak, 1978).
In the occipital cortex, all GABA-ergic neurons are apparently
of non-pyramidal type, the majority of them being aspinous and
sparsely spinous stellate cells (Ribak, 1978; Chronwall and
Wolff, 1978). We have concentrated on the development of these
neurons and we have tried to answer the following questions:

Do all components of the GABA-operated system develop
synchroneously, or does the heterochronicity observed when the
whole cortex is studied (Coyle, 1982) exist even if a small well
defined part of the cortex is used as an experimental model? -
What is the developmental sequence of single components of the
GABA-operated system? - Which components develop in parallel
and which before GABA-ergic synapses? - What may be the role
of non-synaptic GABA during the prenatal and perinatal period?

GENESIS AND POSITIONING OF GABA-ERGIC NEURONS

Neurons in rat visual cortex undergo the last mitosis between
E13 and E21 and begin to migrate along the radial guiding
structures (radial glial cells: Rakic, 1972) to assume the final
positions within the cortical tissue. A majority of neocortical
neurons in layers (L.) II-VI migrate according to a definite
spatio-temporal pattern settling in the deep layers first and in
the superficial layers last (rodents: Angevine and Sidman, 1961;
Berry and Rogers, 1965; primates: Rakic, 1974). However, neurons
in L.I and in the subcortical white matter seem to be positioned
by a different mechanism. This may be the same mechanism which is
continously adding non-pyramidal neurons to all cortical layers
during the whole period of production and migration of cortical

neurons (Rickmann et al., 1977; Chronwall and Wolff, 1978b; Wolff,
1978a). Non-pyramidal neurons are associated with a variety of
transmitters, both excitatory and inhibitory (Emson and Hunt, 1981),
but the majority are GABA-ergic. In order to examine the possibility that GABA-ergic neurons belong to the diffusely (as opposed to
inside-to-outside) positioned subpopulation of non-pyramidal
neurons, we combined the autoradiography based on ^{3}H-thymidine
labeling and GAD-immunocytochemistry. It was demonstrated that
GAD-positive neurons strongly labeled by injection of ^{3}H-thymidine
at E15, E16, E17, E18, E19 and E20, respectively, were scattered
in all cortical layers at P30 (Wolff et al., 1983). Most of the
^{3}H-labeled, GAD-positive neurons were found outside the main,
intensely ^{3}H-labeled, band of GAD-negative (mainly pyramidal)
neurons and there were also a few diffusely positioned, strongly
^{3}H-labeled and GAD-negative neurons. Consequently, it seems that
most of GABA-ergic neurons follow the diffuse, rather then the
inside-to outside mode of positioning and, vice versa, most of the
diffusely positioned neurons are GABA-ergic (Wolff et al., 1983).

FORMATION OF PROCESSES IN GABA-ERGIC NEURONS

    Most of the Golgi-studies in the developing cortex have
focussed either on the differentiation of pyramidal neurons
(e.g. Berry, 1982) or on the postnatal development of non-pyramidal
neurons (e.g. Parnavelas and Uylings, 1980). However, morphological
differentiation of non-pyramidal neurons starts well before birth.
For example, Cajal-Retzius cells in L.I extend about 100 µm long
processes as early as E16. Horizontally oriented neuronal
processes extending along the lower boundary of the cortex have
also been reported (Rickmann et al., 1977) to exist on E16. On
E17 Golgi-stained neurons with horizontally oriented processes
have been observed within the multipolar cortical plate of the rat
cortex and by E20 such neurons have been found in all cortical
layers. During the postnatal period, the number of Golgi-stained
non-pyramidal cells increases and their dendritic trees appear more
elaborate, finally attaining the characteristic stellate shape.
In addition, radially oriented fusiform neurons, not detected
before birth, appear within L.II-V. Some of these cells seem to
be peptidergic excitatory interneurons (see Emson and Hunt, 1981).

    Before birth, axons and axon-like processes in our Golgi-
preparations were limited to a few spiny non-pyramidal neurons. It
is improbable that the absence of axons on other non-pyramidal
neurons is due to selective staining, since in the same preparations
afferent axons and axons with growth cones originating from
pyramidal neurons were relatively frequent (Rickmann et al., 1977;
Chronwall and Wolff, 1978b; Wolff, 1978b). Preliminary observations
on two neurons reconstructed three-dimensionally from serially
resectioned autoradiograms indicated that on E18 in L.I,

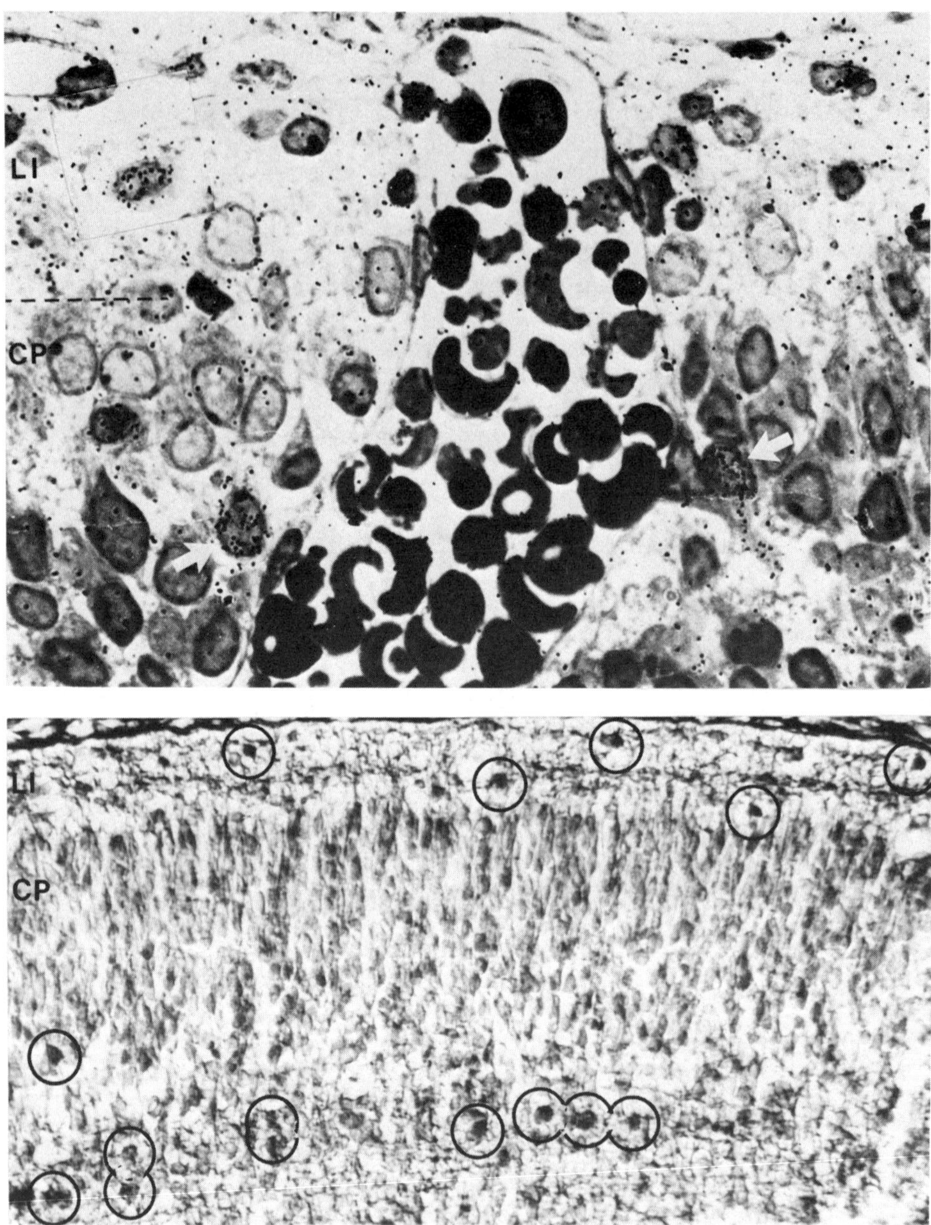

Fig. 1: <u>A</u>: On E16 GABA-accumulating cells are located in L.I (I.) and below the Cortical Plate (CP). Autoradiogram prepared after ^{3}H-GABA superfusion using K2 emulsion and D19 (Ilford) developer, x 1,200. <u>B</u>: On E18 weakly GAD-immunoreactive cells are located in two levels above (I.) and below the cortical plate (CP). Immunohistochemistry using S3-antiserum and PAP method intensified according to Adams (1981), x 200.

GABA-accumulating neurons may not yet have grown axons but have
well developed dendrites forming and receiving dendro-dendritic
synapses. In the same sections, a neuron which had not accumulated
GABA did grow an axon as well as two main dendrites (M. Rickmann,
unpublished observations). This may mean that the development and
differentiation of dendrites of non-pyramidal neurons start well
before birth, while in contrast, the growth of axons and the
formation of axonal synapses may not occur, at least not on
GABA-ergic neurons, until birth or even later. Similarly, GABA-
accumulating neurons do not seem to receive axosomatic synapses
until shortly before birth (Chronwall and Wolff, 1980) although
synapses have been found on dendrites several days earlier.

GABA UPTAKE

Uptake of GABA is one of the most important neurochemical
markers of GABA-ergic inhibition. The role of GABA uptake in the
adult nervous tissue is probably twofold: (i) to keep the extracellular concentration of GABA low, thus preventing uncontroled
hyperpolarization of neuronal membranes and, (ii) to replenish
intracellular (synaptic) stores of GABA.

In the adult cortex GABA is accumulated by perikarya and
synaptic terminals of putative GABA-ergic neurons and, to a
smaller extent, by a subpopulation of glial cells (Hökfelt and
Ljungdahl, 1971; Wolff and Chronwall, 1978a). Some meningeal
cells, pericytes and a few endothelial cells of intracerebral
blood vessels also take up GABA, but the contribution of these
structures to the uptake of GABA in the adult visual cortex is
relatively minor.

Ontogenetic development of GABA uptake was studied by two
methods. Morphology and distribution of GABA-accumulating
structures was determined autoradiographically in cortical tissue
superfused or injected with ³H-GABA in vivo (Wolff and Chronwall,
1978a; Chronwall and Wolff, 1980), while quantitative and kinetic
aspects were studied in tissue slices in vitro (Iversen and Neal,
1967).

Development of GABA-accumulating neurons and glial cells,
as visualized by autoradiography follows a characteristic
spatio-temporal pattern (Chronwall and Wolff, 1980). The first
³H-labeled cell bodies appeared on embryonic day 16 (E16) in
lamina 1 (L.I) and in the multipolar part of the cortical plate,
including subcortical levels (i.e. presumptive L.IV-L.VI and white
matter, Fig. 1A). On E18, isolated ³H-labeled cells existed in all
cortical layers and on E20 clusters of ³H-labeled cells appeared
within the bipolar cortical plate (presumptive L.II-III/IV).
By postnatal days (P) 1 to 3 GABA-accumulating cells were evenly
distributed throughout the cortex and the adult numerical density

DISTRIBUTION OF GABA ACCUMULATING NEURONS

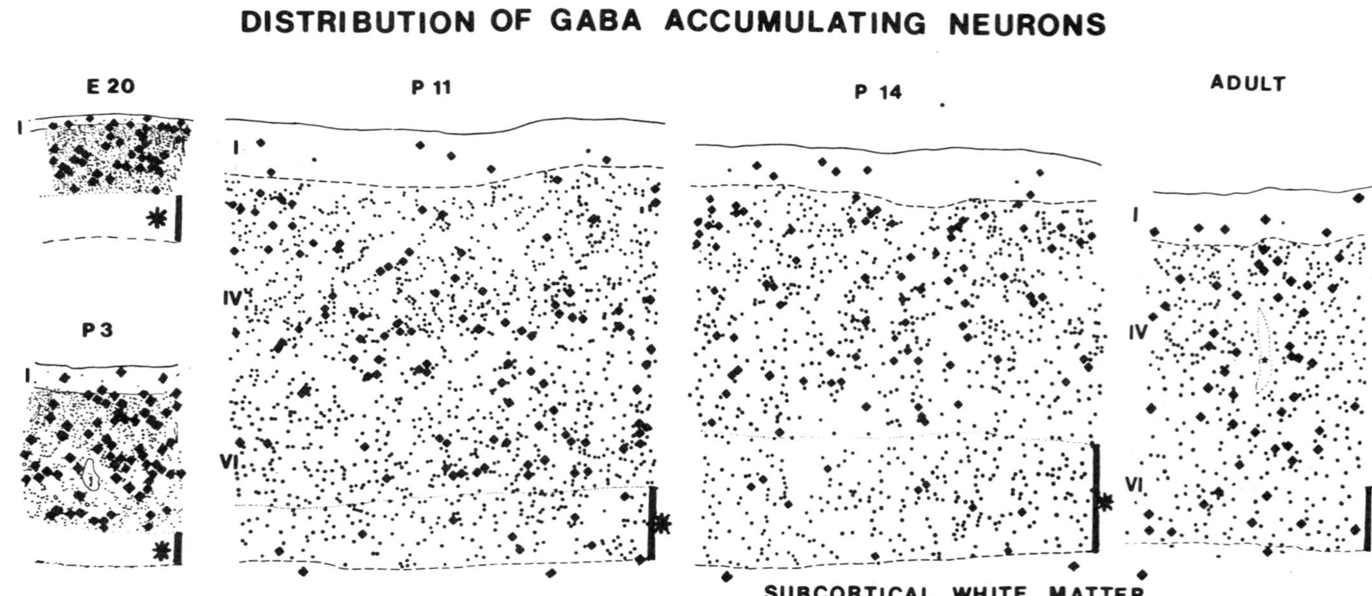

Fig. 2: Plots of autoradiograms: GABA accumulating neurons (large dots) are intermingled with unlabeled neurons (small dots) in all layers. In some areas (*) the labeling was missing (E20, P3) or incomplete. E = embryonic, P = postnatal days. The cortical thickness shows an overshoot at about P10.

was reached by P10 (Fig. 2). Since the volume of the cortex
continues to grow after P10 and more than doubles between P8 and
P55 (Bähr and Wolff, in press), the total number of GABA-accumulating neurons must also increase. Consequently, after this the
expression of GABA uptake in neurons may be related to the growth
of the neuropil.

Cytological characterization of GABA-accumulating cells was
done by electron microscopy in resectioned autoradiograms. From
E17, GABA-accumulating neurons could be positively distinguished
from GABA-accumulating glial cells. Initially, ^{3}H-GABA-accumulating
neurons represented a relatively small fraction of all neurons in
L.I but after birth great majority of L.I neurons became labeled
(Tab. 1). This may be due to the disappearance of non-GABA-ergic
Cajal-Retzius neurons although a delayed expression of GABA uptake
in some neurons cannot be ruled out.

Table 1: GABA-accumulating neurons as percentage of all neurons

layers	E20	P3	P11	P14	P22	adult
I	15	45	75	78	75	75
V	5	10	13	12	12	12
I-VI	5	10	9	10	9	9

In L.I and L.V the fractions of GABA-accumulating neurons increase until P11, while the adult value is reached by P3 in the other layers.

In the rat, the first neurons enter the anlage of the occipital
cortex about E14 and last neurons arrive perinatally (Raedler and
Sievers, 1975; Rickmann et al., 1977). The fact that the number
of cortical neurons accumulating ^{3}H-GABA continues increasing well
after the birth (Tab. 1), would seem to indicate that the
expression of GABA uptake in neurons does not follow a rigid time
course.

In contrast to earlier views, non-radial glial cells are now
thought to develop within the cerebral cortex starting at early
developmental stages (Rickmann and Wolff, 1977). Many of the
embryonic non-radial (and some radial) glial cells accumulate GABA,
at least at some stage during the differentiation (Chronwall and
Wolff, 1980). Non-radial and radial glial cells form a spongy
framework which may be important for the migration and positioning
of glioblasts and non-pyramidal neurons (Wolff et al., 1978;
Rickmann and Wolff, in press).

Neurochemistry and pharmacology of GABA uptake was described
in vitro using brain slices and synaptosomes (Iversen and Johnston,
1971). The uptake process is mediated by an active, highly
substrate-specific, saturable transport system with an apparent

K_m in low micromolar region ("high affinity uptake"). The driving
force of the active transport seems to be the gradient of Na^+
concentration across the cellular membrane (Martin, 1976), although
the process of homoexchange may also contribute to the transport of
^{3}H-GABA against an apparent concentration gradient in the
experiments in vitro (Levi and Raiteri, 1974; Fagg and Lane, 1979)
and perhaps in vivo.

A weak uptake of GABA is detectable in the slices and
homogenates of neonatal cortical tissue (Johnston and Davies, 1974;
Coyle and Enna, 1976). Uptake of GABA in immature cortical slices
is more sensitive to the inhibition by β-alanine (putative marker
for GABA uptake system in glial cells, Iversen and Kelly, 1976)
than the uptake of GABA in adult cortex (Table 2). According to
Wong and McGeer (1981) uptake of ^{3}H-β-alanine remains almost
constant during the postnatal development but uptake of ^{3}H-cis-3-
aminocyclohexane-1-carboxylic acid (ACHC, putative marker for the
neuronal uptake of GABA; Neal and Bowery, 1977) increases several-
fold between P0 and P30. However, quantitative calculations based
on such data may not be reliable, since the substrate specificity
of GABA uptake in immature glial cells and neurons may differ from
that of GABA uptake in adult cortical cells (Balcar et al., 1979;
Balcar and Hauser, 1982).

In the occipital cortex the maximum rate of uptake (V_{max}) at
P0 is about 2 % of the adult value. The most rapid increase in
uptake capacity takes place during the second postnatal week
(Fig. 3A) almost coinciding with the rapid increase of K^+-stimul-
ated release of GABA (Balcar et al., 1983), but preceding the
most rapid phase in the development of type 2 synapses by several
days (Fig. 3E). This suggests that during the development of the
occipital cortex a substantial proportion of GABA uptake is
localized in non-synaptic structures such as glial cells, neuronal
dendrites and free axonal varicosities which have not yet formed
synaptic contact.

Table 2: Uptake of ^{3}H-GABA in vitro

	P0	P9	Adult
β-Alanine	40 ± 4*	14 ± 11	9 ± 5
L-2,4-Diaminobutyrate	51 ± 10*	54 ± 3*	57 ± 7*

Effects of 100 µM β-alanine and L-2,4-diaminobutyrate
on the uptake of 1 µM ^{3}H-GABA by slices of cortical
tissue at various stages of postnatal development.
Results expressed as % inhibition (Mean $\pm$ S.D. of
4 determinations.
* $P < 0.02$ by Student's t-test

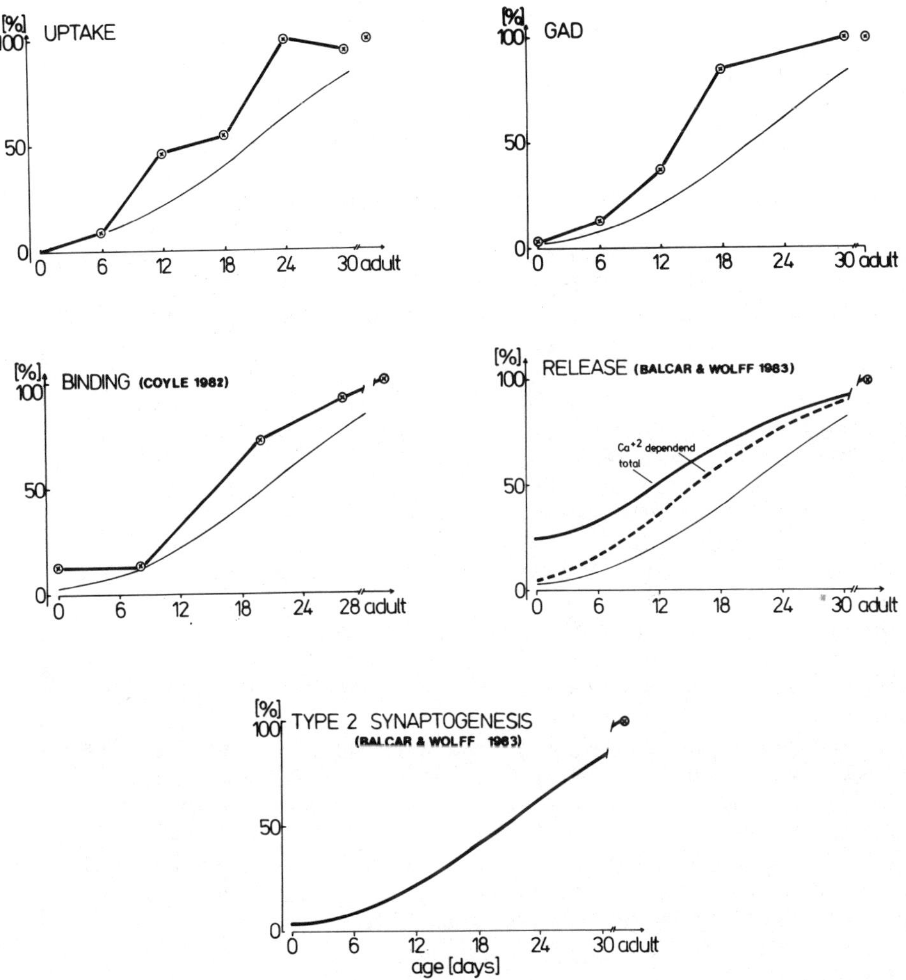

Fig. 3: Time courses of GABA-ergic neurochemical markers (A-D) and the formation of type 2 synapses (E). Note that type 2 synaptogenesis (thin lines in A-D) lags behind the other parameters.

GABA RELEASE

The K^+-stimulated, Ca^{+2}-dependent release of ^3H-GABA from
brain slices, an experimental model of synaptic release of GABA
(Katz et al., 1969;Srinivasan et al., 1969) has been shown to
exist in the slices of rat cortex even at PO (Davies et al., 1975;
Balcar et al., 1983). However, the synaptic localization of
GABA release in the developing cortex has recently been questioned
for at least two reasons. Firstly, the release mechanism is less
Ca^{+2}-dependent in the newborn than in the adult cortex,
especially at lower concentrations of K^+ (Davies et al., 1975;
Balcar et al., 1983). Secondly, while the Ca^{+2}-independent
portion of the release remains about constant during the
development, the Ca^{+2}-dependent part increases rapidly,
especially during the second postnatal week and this does not
correlate with the fast phase of the development of type 2
synapses which arrives about a week later (Fig. 3B,E). Accordingly,
at least a part of the release mechanism is likely to be localized
in non-synaptic structures. The Ca^{+2}-independent part is likely to
represent the release from glial cells (Johnston, 1977), but the
Ca^{+2}-dependent release during the first and second postnatal week
may be associated with neuronal structures. In cultured neonatal
cortical GABA-ergic neurons, there was no stimulated release of
GABA during the first week in culture although GABA, GAD-activity
and GABA uptake were all present (Hauser et al., 1980; Balcar and
Hauser, 1982). The release of GABA then increased at least 10 days
earlier than either GABA uptake or GAD activity. Although some
synapses may have been present even at the 5th day in culture
(Hauser and Heid, 1978), the only important morphological change
accompanying the rapid development of the release mechanism was a
major increase in the complexity of the neurites. In other studies,
Redburn et al. (1978) showed that synaptosomes prepared from
developing rat cortex do not release GABA in Ca^{+2}-dependent manner
during the first postnatal week and even at the end of the third
postnatal week the potency of Ca^{+2}-dependent GABA-releasing
mechanism was less than 50 % of the adult value. Consequently, it
seems that most of the K^+-stimulated, Ca^{+2}-dependent release of
GABA observed during the first and second postnatal week comes
either from neuronal perikarya and dendrites or from free GABA-ergic
axons which have not yet formed synapses.

GABA SYNTHESIS

The decarboxylation of L-glutamate catalysed by L-glutamate-
1-carboxylyase(EC 4.1.1.15, glutamate decarboxylase GAD) is the
major source of synaptic GABA in the adult neocortex (Ribak, 1978).
In the developing tissue, an alternative synthetic pathway originating from putrescine has also been considered but found to be
of only minor importance (Seiler and Sahar, 1983). In the view
of intimate association between the GAD-catalysed synthesis and

the inhibitory action of GABA (for ref. see e.g. Horton, 1980)
studying of GAD at various postnatal stages seems crucial to the
understanding of the ontogenetic development of GABA-ergic
inhibition in the neocortex.

We approached the problem by <u>two different methodologies.</u>
Topography and cytology of GAD have been studied by immunohistochemical techniques in which the cells and cell processes containing GAD were labeled by the S3-antiserum (Oertel et al., 1982)
and visualized by the peroxidase-antiperoxidase (PAP) procedure
(Wolff et al., in press) while the quantitative, kinetic and neurochemical aspects of GAD have been analysed in vitro by measuring
production of $^{14}CO_2$ from $[1-^{14}C]$ L-glutamate in homogenates of
neocortical tissue (Wilson et al., 1972).

Using Nomarski optics or heavy metal intensification,
<u>GAD-like immunoreactivity</u> has been visualized as early as on E18
(Adams, 1981; Gallyas et al., 1982) and it is not impossible that
more sensitive methods would reveal very small amounts of GAD at
even earlier stages. Before birth, most of the GAD-positive cell
bodies were stained weakly and were distributed mainly within two
horizontal layers: (i) in L.I about 50 % of the neurons were
GAD-positive and (ii) in the subplate (i.e. multipolar cortical
plate plus intermediate zone, which is the presumptive white
matter) where only isolated cells were labeled (Fig. 1B). During
the perinatal period this bimodal distribution changed into a
three-tiered pattern. GAD-like immunoreactivity was visible
(i) in cells located in L.I (including the superficial part of
L.II), (ii) in some weakly immunoreactive cells and cell
processes in the transition zone between bipolar and multipolar
cortical plate (prospective L.V/IV) and (iii) in cells of L.VI
and the presumptive white matter. At about the time of birth, long
beaded GAD-positive processes (possibly axons) appeared and grew
in number in both L.I and along the lower boundary of the cortex.
During the second postnatal week, the numerical density of weakly
stained neuronal cell bodies reached maximum values in all
layers (Fig. 4) and GAD-positive axon varicosities multiplied,
mainly in layers I, V and VI. GAD-positive processes formed
perisomatic baskets around pyramidal neurons first in L.V and
L.VI and a few days later also in L.III and L.II. During the 3rd
postnatal week, GAD-positive cell bodies reached adult numerical
densities in all cortical layers while the density of GAD-positive
axon varicosities reached approximately adult values about a week
later. Furthermore, during the 4th postnatal week the axonal
varicosities in the primary and secondary visual areas,
respectively, developed different laminar patterns. In area 17 the
density of GAD-positive axon terminals and varicosities decreased
in L.VI, so that the supragranular layers contained on average
a higher density of GAD-positive structures than the infragranular
layers and GAD-positive perisomatic "baskets" became more

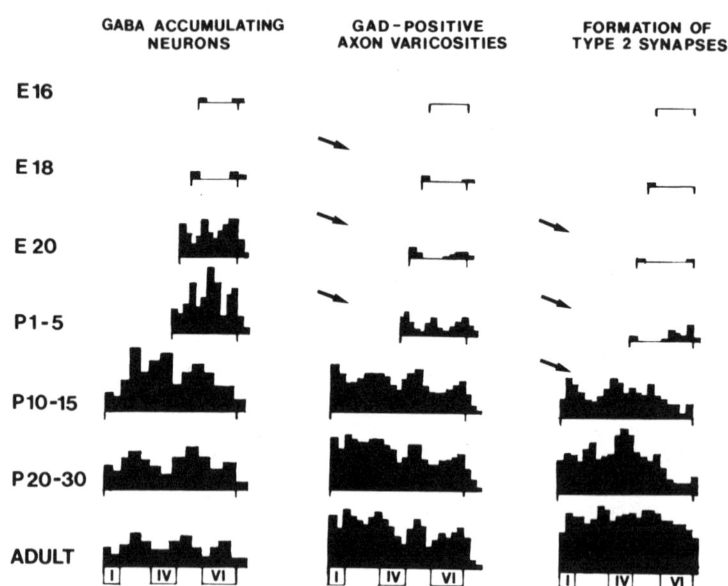

Fig. 4: Development of laminar distribution patterns of GABA-accumulating neurons, GAD-positive axon varicosities, and type 2 synapses in the visual cortex of rat. The numerical densities of the respective structures are adapted to expose the laminar patterns but are not in absolute units. Note that similar laminar patterns are reached at different developmental stages (arrows).

numerous and consisted of larger axon varicosities. In some cases
both perikarya and adjoining perisomatic baskets were labeled.
There were no systematic changes in the distribution and density
of GAD-positive structures after the 4th postnatal week (further
details see Wolff et al., in press). Relatively large variations
in GAD-like immunoreactivity in neuronal somata and processes
were observed, even in the adult visual cortex. Such phenomenon
may be related to functional states of the cortex (e.g. a recent
"history" of excitation) rather than to different stages of
maturation.

The experiments _in vitro_ demonstrated that the occipital
cortex, like the parietal cortex (Coyle and Enna, 1976), is
capable of decarboxylating L-glutamate in a specific manner even
at P0. However, the total activity of GAD was very low during the
first postnatal week (ca. 1-3 % of the adult value) and the
structural specificity of the enzyme seemed to be different from
that in the adult cortex (Table 3, Taberner et al., 1977).

Table 3: Sensitivity of GAD activity to inhibitors

(A)	P0	adult
L-glutamate-γ-hydroxamate	72 $\pm$ 5	97 $\pm$ 1*
L-glutamate-γ-hydrazide	78 $\pm$ 2	94 $\pm$ 2*
L-aspartate-β-hydroxamate	N.S.	76 $\pm$ 6*
D-aspartate-β-hydroxamate	30 $\pm$ 5	75 $\pm$ 6*
(B)		
Pyruvate 10 mM	N.S.	N.S.
DTNB 0.01 mM	N.S.	N.S.
DTNB 1.0 mM	98 $\pm$ 2	98 $\pm$ 1
NaCl 4 mM	81 $\pm$ 4	78 $\pm$ 5

The figures represent percent inhibition of GAD activity
expressed as mean $\pm$ S.D. of four independent determinations.
The control values were 0.22 $\pm$ 0.03 (6 determinations) and
4.02 $\pm$ 0.47 (8 determinations) for P0 and adult, respectively.
Not significant (N.S.) signifies that the mean was less than
10 % different from the controls and P $\geqq$ 0.02. Concentration
of L-glutamate was 1 mM and concentrations of test substances
were, unless otherwise stated, 5 mM.

*P < 0.02 by Student's t-test compared to the value at P0

GAD activity increased sharply during the second postnatal week
and reached adult level at approximately P20 (Fig. 3C) although
biochemical characteristics suggest that GAD at P0 is closer to
the neuron-localized enzyme in the adult brain rather than to the
non-neuronal form of GAD described e.g. in heart muscle (Wu, 1979,
Table 3B). A comparison of the time course of development of GAD
activity with that of type 2 synapses implies that a position of
GAD activity especially during the 3rd and 4th postnatal week is
localized in non-synaptic, possibly in non-neuronal structures.

GABA RECEPTORS

A large number of pharmacological and neurochemical studies
indicates that Na^+-independent binding of GABA to neuronal
membranes in vitro is a satisfactory experimental model of
GABA-specific recognition sites associated with postsynaptic
receptors of GABA (Costa et al., 1981).

In the cerebral cortex, Na^+-independent binding of GABA is
present at birth (ca. 10 % of the adult value) and remains
approximately constant until at least P7 (Coyle, 1982). During
the second and third postnatal week, the binding of GABA in
cortex increases to about 80 % of the adult value (Fig. 3D),
approximately coinciding with, or somewhat preceding, the development of type 2 synapses (Fig. 3E). This suggests an interaction
between the formation of GABA-ergic synaptic contacts and the
sensitivity of the postsynaptic membrane to GABA (or vice versa).
In this context, it is interesting to note that GABA also induces
GABA-binding sites in cultured neurons (Meier et al., 1983). The
time course of development of GABA binding sites has not yet been
determined in the visual cortex. Slight differences from the data
mentioned above (whole cortex) cannot be ruled out.

FORMATION OF GABA-ERGIC SYNAPTIC CONTACTS

Studying synaptogenesis necessarily involves determination of
the numbers of synapses per unit of tissue at various stages of
ontogenetic development. Direct quantitative estimations (i.e.
counting of GABA-ergic synapses visualized by GAD-immunocytochemistry in corticle preparations) is practically impossible due
to numerous methodological difficulties. Consequently, approxim-
ations through an indirect approach have to be done. For example,
a simple solution would be to identify GABA-ergic synapses
morphologically, i.e. by a characteristic structural feature.

In a typical cortex preparation there are synapses of many
different structures. This is partly caused by differences in the
structure of pre- and/or postsynaptic elements or by variations
of their orientation relative to the plane of sectioning.
Fortunately, on the basis of the distribution of the electron-dense

Table 4: Structure of GAD-positive synapses

age (weeks p.n.)	n	type 1 synapses	intermediate forms	type 2 synapses
2	305	3.6 %	1.6 %	94.7 %
3	202	12.4 %	2.5 %	85.1 %
6	96	1.0 %	4.2 %	94.8 %
7	196	2.6 %	1.0 %	96.4 %

material in the vicinity of the synaptic cleft, it is possible to classify the synapses either as asymmetrical (type 1) and symmetrical (type 2) (Gray, 1959; Colonnier, 1968). The size and shape of the transmitter-storing synaptic vesicles, as well as other criteria, suggested that the synapses of type 1 and 2 represent excitatory and inhibitory synapses, respectively (for ref. see Uchizono, 1975). This may be an oversimplification, and since there are intermediate types of synapses in various regions of the CNS, the structure and function of synapses are probably not directly related to each other. However, in the occipital cortex of adult rats, GABA-ergic neurons, as visualized by GAD-immunocytochemistry (Ribak, 1978), form almost exclusively synapses of type 2 (Table 4). The converse also seems to be true, namely the synapses of type 2 in the visual cortex are mostly GABA-ergic. This has been confirmed by electron microscopic analysis of the autoradiograms of ^{3}H-GABA-labeled cortical tissue (Wolff and Chronwall, 1982) and in our immunocytochemical material using GAD antibodies (Fig. 5).

During the ontogenetic development, symmetric synapses (type 2, inhibitory) first appear in L.I on E17 and in the deepest cortical layers on E18 or E19. At about the time of birth, additional symmetric synapses are found in L.V and their development continues in all cortical layers. A fourth maximum in the numerical density of symmetric synapses (apart from the maxima in layers I, V, deep VI) appears in L.IV during the second postnatal week. Finally, by the end of the first postnatal month, symmetric synapses become almost evenly distributed throughout the cortex.

The total numerical density of symmetric synapses in the occipital cortex remains low until the end of the first postnatal week. Then it begins to increase and reaches about 50 % of the adult value during the third week. The time course of the formation of symmetric synapses varies somewhat from layer to layer and, also, the axosomatic synapses develop 1 to 2 days ahead of axodendritic ones (Bähr and Wolff, submitted). The

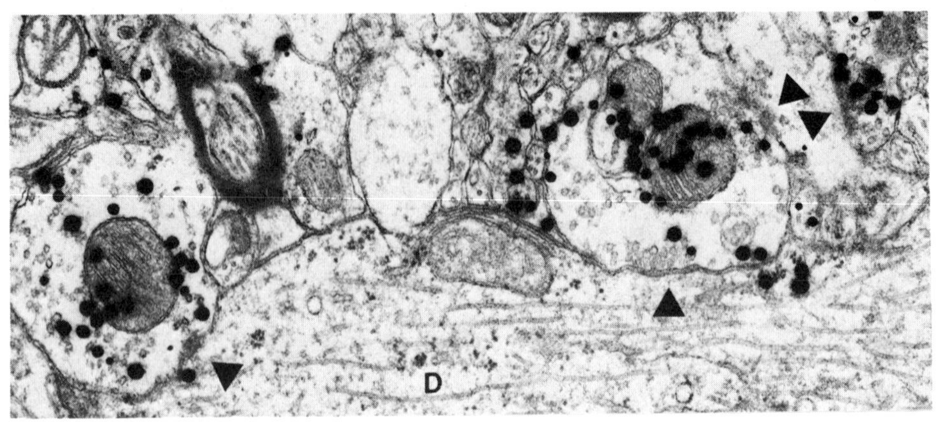

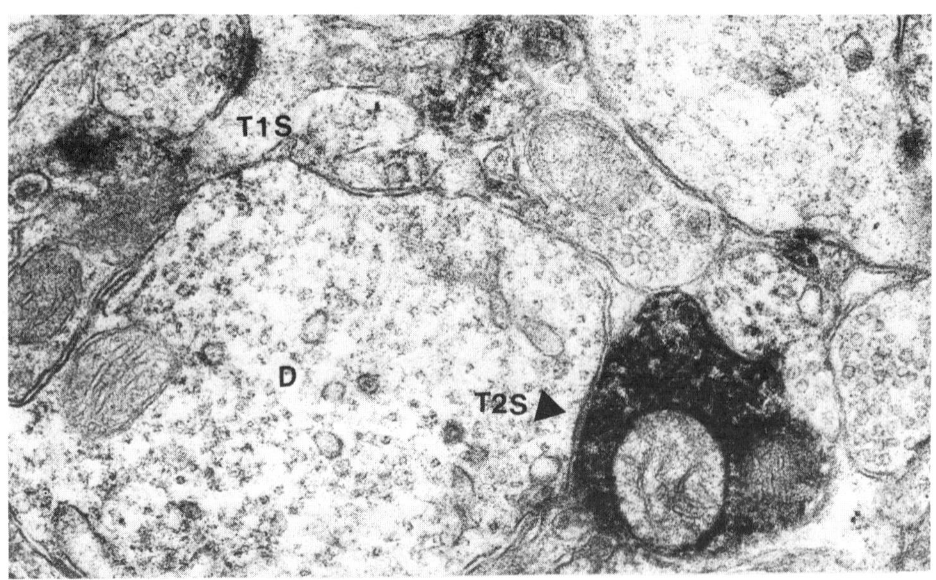

Fig. 5: GABA-ergic axon terminals were identified by two techniques, ^{3}H-GABA accumulation followed by autoradiography using L4 emulsion (Ilford) and Amidol (Merck) developer (A) and GAD-like immunoreactivity using S3-antiserum and the PAP-method (B). Note that in both preparations GABA-ergic synapses form type 2 contacts (T2S, arrow heads). A: x 30,000, B: 40,000 D: dendrite, T1S = type 1 synapse.

230

total numerical density of symmetric synapses is about 15 % short
of the adult value at P30 and stabilizes only during the second
postnatal month (Balcar et al., 1983; Fig 3E). Adult values
are reached about 2 months after birth. This indicates that in
the rat visual cortex the synaptogenesis of type 2 synapses lags
behind that of type 1 (Wolff, 1978).

GABA-ergic synapses may undergo structural transformations
during the reactive synaptogenesis. For example, GAD-positive
type 1 synapses appear in the adult superior colliculus following
de-afferentation (Houser et al., 1983). Consequently, we felt
that the possibility of a transformation of the synaptic
structure (e.g. from type 1 to type 2) should be also examined
during the primary synaptogenesis (i.e. the synaptogenesis during
the ontogenetic development). Sections prepared by Vibratome from
cortical tissue at various stages of postnatal development were
first stained by GAD-antiserum to visualize GABA-ergic structures,
then embedded in Epon 812, re-sectioned perpendicularly to the
original plane of sectioning and examined by electron microscopy.
The synapses formed by the presynaptic endings cut open by the
first sectioning were classified according to morphological
criteria and the presence or absence of GAD-like immunoreaction
product. Exclusion of the structures left intact by the first
sectioning eliminated the synaptic endings which could contain
GAD, but were not labeled since the antibodies had not penetrated
the intracellular space. Preliminary results (Table 4) indicated
that most of the GAD-positive presynaptic terminals formed indeed
type 2 synapses even at early stages of postnatal development. The
proportion of GAD-positive synapses of type 1 remained within
1 to 4 % of the total (perhaps with the exception of P21). This
indicates that the numerical densities of the symmetric (type 2)
synapses represent a good approximation of numerical densities
of GABA-ergic synapses during the ontogenetic development.

CONCLUSIONS

A synopsis of the reviewed data suggests that the development
of GABA-ergic neurons in the visual cortex proceeds in several
steps. At least four separate phases can be distinguished:
(1)   During the last mitosis and throughout the migration and
positioning, the precursors of GABA-ergic neurons express neither
the uptake of GABA nor GAD, detectable by autoradiography, immunocytochemistry or biochemically in vitro. Hence, it is not known
whether there are clones of (pre)neurons destined to become GABA-
ergic or, rather, the nature of the synaptic transmitter is
determined after the neurons have settled in permanent positions.
(2)   One to several days after the arrival in the definite
positions within the occipital cortex, presumptive GABA-ergic
neurons start to form dendrites. At this stage, neurons contain
small but detectable amounts of GAD and the periarya and

dendrites can be <u>labeled by</u> ^{3}H-GABA. Furthermore, the neurons form
apparent dendro-dendritic synapses and begin to receive input from
axons of unknown origin through asymmetric synapses. This phase
lasts several days in the presumptive GABA-ergic neurons which
start to differentiate before birth but it is not clear whether,
in the neurons starting to differentiate postnatally, this phase
can be separated from the following one.
(3)   Non-pyramidal, aspinous neurons start to develop <u>axons</u>
between E18 and about the end of the first postnatal week. These
axons form increasing numbers of varicosities and potential
presynaptic terminals, both of which accumulate GABA and contain
GAD. Data from neurochemical experiments in vitro suggest that
under the conditions of increased extracellular concentration of
K^+, such terminals may release GABA in a Ca^{+2}-dependent (i.e.
synaptic-like) manner. A comparison of the laminar distribution
of GAD-positive axon varicosities with that of type 2 synapses
suggests that many of the axon varicosities remain free for
several days before making synaptic contacts.
(4)   Formation of <u>inhibitory (type 2) synapses</u> completes the
differentiation of GABA-ergic neurons. Most of the GABA-ergic
synapses (75 %) develop between the 2nd and 8th postnatal week,
i.e. after thalamic afferents have made synapses and after the
eyes have opened. However, neurochemical markers, normally
associated with functioning GABA-ergic synapses, reach about 50 %
of the adult values before that period, clearly preceding the
formation of type 2 synapses (Fig. 3A-3F). Consequently,
synthesis, release and accumulation of GABA are not dependent on
the existence of synaptic contacts.

The same gradated sequence seems to be followed by GABA-ergic
neurons in all cortical layers (Fig. 4).

Data discussed so far, demonstrate that there is a period in
the development of GABA-ergic neurons when the synthesis, uptake
and release of GABA exist, but the GABA-ergic synaptic contacts
are relatively sparse (phase 2-3). This raises an interesting
question - does GABA, or do GABA-ergic structures have a
specific role during the stages of differentiation preceding the
synaptogenesis? Several lines of evidence suggest that GABA may
be involved in other than synaptic mechanisms: (i) Glial cells
may release GABA in response to increased concentrations of K^+
(e.g. as a consequence of hyperactivity of neighbouring neurons,
Bowery and Brown, 1972). (ii) There are examples of glia-localised
GAD (subcommissural organ: Weissmann-Nanopoulos et al., 1983;
neocortex: Böttcher, Stuke, Balcar and Wolff, unpublished). Hence,
GABA may not originate exclusively in GABA-ergic neurons. (iii)
GABA-sensitive receptors exist even on neurons with no GABA-ergic
synaptic input, e.g. sympathetic ganglia (Adams and Brown, 1975).

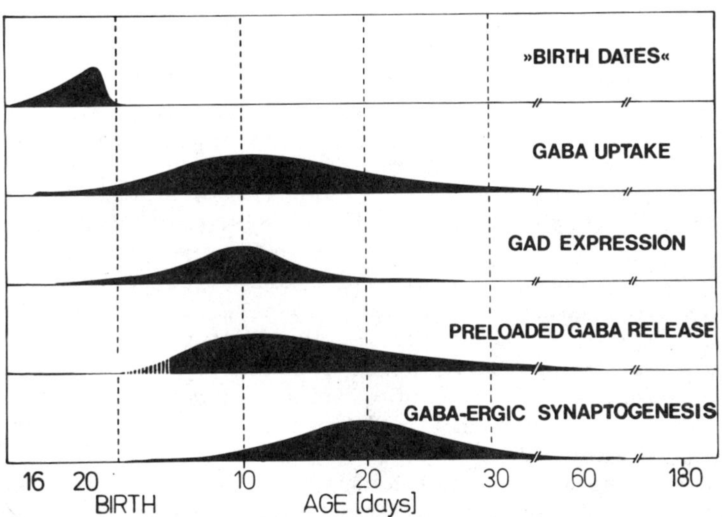

Fig. 6: Synopsis of the formation rates (first derivations of time courses) of non-pyramidal neurons ("birth dates") and some parameters of the GABA operated system. Note time shifts between "birth dates" and the GABA-ergic parameters and between the latter and GABA-ergic synaptogenesis.

In addition, GABA modifies characteristics of GABA-binding
sites in cultured neurons (Meier et al., 1983) and induces a
formation of free postsynaptic densities on the neurons in
sympathetic ganglia in situ (Wolff et al., 1979). In the latter
preparation this effect facilitates formation of additional
excitatory (cholinergic) synaptic input to the GABA-treated
neurons (Wolff et al., 1981; Dames and Feher, unpublished). On
the basis of these observations a theory has been proposed
according to which an excess of inhibition/excitation induces
(or facilitates) formation of excitatory/inhibitory synaptic
input (Wolff, 1981a,1981b; Wolff and Wagner, 1983). It is
possible that, during the development, non-synaptic (e.g. glial)
compartment stores and releases GABA which provides the
inhibition regulating the formation of excitatory synapses. In
this context, it is interesting to note that GABA-accumulating
cells appear early in the ontogenetic development (Rickmann and
Wolff, in press) and some of them may transiently receive
synaptic input (for ref. see Wolff et al., 1979).

In conclusion, differentiation of GABA-ergic neurons in the
rat visual cortex follows a characteristic sequence involving
(i) migration and positioning, (ii) appearance of GABA-accumulating perikarya and dendrites, (iii) growth of axons and axonal
varicosities, and (iv) formation of synaptic contacts. There are
indications that during (ii) and (iii) GABA, stored predominantly
in a non-synaptic compartment, has a specific role in the process
of differentiation. Previous studies in the superior cervical
ganglion suggested that GABA, under certain experimental
conditions, exerts a morphogenetic effect on neurons. A synopsis
of the data discussed in this review implies that a similar
mechanism involving GABA may regulate interactions among
developing neurons in the visual cortex, thus influencing the
development and stabilization of synaptic contacts.

REFERENCES

Adams, P.R., and Brown, D.A., 1975, Action of γ-aminobutyric acid
    on the sympathetic ganglion cells, J. Physiol., London,
    250: 85-120.
Adams, J.C., 1981, Heavy metal intensification of DAB-based HRP
    reaction product, J. Histochem. Cytochem., 19: 755.
Anvengine, J.B., Jr., and Sidman, R.L., 1961, Autoradiographic
    study of cell migration during histogenesis of cerebral
    cortex in the mouse, Nature, 192: 766-768.
Bähr, S. and Wolff, J.R., Postnatal development and loss of
    axosomatic synapses in the rat visual cortex. Morphogenesis
    and quantitative evaluation of type 1 and type 2 synapses
    (submitted).

Balcar, V.J., Mark, J., Borg, J., and Mandel, P., 1979, High
affinity uptake of γ-aminobutyric acid in cultured glial
and neuronal cells, Neurochem. Res., 4: 339-354.

Balcar, V.J., and Hauser, K.L., 1982, Development of uptake of
γ-aminobutyrate in cultured neurons, 12th Intern. Congr.
Biochem. (Abstract), 95.

Balcar, V.J., Dammasch, I., and Wolff, J.R., 1983, Is there a
non-synaptic component in the K⁺-stimulated release of
GABA in the developing rat cortex? Dev. Brain Res.,
10: 309-311.

Berry, M., and Rogers, A.W., 1965, The migration of neuroblasts
in the developing cerebral cortex, J. Anat., 99: 691-709.

Berry, M., 1982, Cellular differentiation: Development of
dendritic arborizations under normal and experimentally
altered conditions, in: "Development and Modifiability of
the Cerebral Cortex", P. Rakic and P.S. Goldman-Rakic,
eds., Neurosci. Res. Progr. Bull., 20: 451-461.

Bowery, N.G., and Brown, D.A., 1972, γ-aminobutyric acid uptake
by sympathetic ganglia, Nature, 238: 89-91.

Chronwall, B.M., and Wolff, J.R., 1978a, Classification and
location of neurons taking up ³H-GABA in the visual cortex
of rats, in: "Amino Acids as Chemical Transmitters", Plenum
Press, New York, pp. 297-303.

Chronwall, B.M., and Wolff, J.R., 1978b, Aspects on the development
on non-pyramidal neurons in the neocortex of rat, Zoon,
6: 145-148.

Chronwall, B.M., and Wolff, J.R., 1980, Prenatal and postnatal
development of GABA-accumulating cells in the occipital
cortex of rat, J. Comp. Neurol., 190: 187-208.

Costa, E., Di Chiara, G., and Gessa, G.L., ed., 1981, "GABA and
Benzodiazepine Receptors", Adv. Biochem. Pharmacol., 26,
Raven Press, New York.

Coyle, J.T., and Enna, S.J., 1976, Neurochemical aspects of the
ontogenesis of GABA-ergic neurons in the rat brain, Brain
Res., 111: 119-133.

Coyle, J.T., 1982, Development of neurotransmitters in the neo-
cortex, in: "Development and Modifiability of the Cerebral
Cortex", P. Rakic and P.S. Goldman-Rakic, eds., Neurosci.
Res. Progr. Bull. 20: 479-492.

Davies, L.P., Johnston, G.A.R., and Stephanson, A.L., 1975,
Postnatal changes in the potassium-stimulated, calcium-
dependent release of radioactive GABA and glycine from
slices of rat central nervous tissue, J. Neurochem.,
387-392.

Emson, P.C., and Hunt, S.P., 1981, Anatomical chemistry of the
cerebral cortex, in: "The Organization of the Cerebral
Cortex", F.O. Schmitt, F.G. Worden, G. Adelmann, S.G.
Dennis, eds., MIT-Press, Cambridge, Massachusetts, and
London, England, pp. 325-345.

Fagg, G.E., and Lane,J.D., 1979, The uptake and release of
     putative amino acid transmitters, <u>Neuroscience</u>, 4:
     1015-1036.
Gallyas, F., Görcs, T., and Merchenthal, I., 1982, High-grade
     intensification of the end product of the diaminobenzidine
     reaction for peroxidase histochemistry, <u>J. Histochem.</u>
     <u>Cytochem.</u>, 30: 183-184.
Hauser, K.L., and Heid, J., 1978, Morphology and biochemistry of
     rat cortical neurons in dissociated cell culture, <u>Proc.</u>
     <u>Eur. Soc. Neurochem.</u>, 1: 503.
Hauser, K.L., Balcar, V.J., and Bernasconi, R., 1980, Development
     of GABA neurons in dissociated cell culture of rat
     cerebral cortex, <u>in</u>: "GABA Neurotransmission", H. Lal,
     ed., Brain Res. Bull. 5, suppl. 2: 37-41.
His, W., 1904, Die Entwicklung des menschlichen Gehirns während
     der ersten Monate, Hirzel Verlag, Leipzig.
Hökfelt, T., and Ljungdahl, A., 1972, Autoradiographic identific-
     action of cerebral and cerebellar cortical neurons accum-
     ulating labeled gamma-aminobutyric acid (^{3}H-GABA). <u>Exp.</u>
     <u>Brain Res.</u>, 14: 354-362.
Horton, R.W., 1980, GABA and seizures induced by inhibition of
     glutamic acid decarboxylase, <u>in</u>: GABA Neurotransmission",
     H. Lal, ed., <u>Brain Res. Bull.</u> 5, suppl. 2: 605-608.
Houser, C.R., Lee, M., and Vaughn, J.E., 1983, Immunocytochemical
     localization of glutamic acid decarboxylase in normal and
     de-afferented superior colliculus: Evidence for reorgan-
     ization of γ-aminobutyric acid synapses. <u>J. Neurosci.</u>
     3: 2030-2042.
Iversen, L.L., and Neal, M.J., 1968, The uptake of ^{3}H-GABA by
     slices of rat cerebral cortex, <u>J. Neurochem.</u>, 15:
     1141-1149.
Iversen, L.L., and Johnston, G.A.R., 1971, GABA-uptake in rat
     central nervous system: Comparison of uptake in slices
     and homogenates and the effects of some inhibitors.
     <u>J. Neurochem.</u> 18: 1939-1950.
Iversen, L.L., and Kelly,J.S.,1975, Uptake and metabolism of
     γ-aminobutyric acid by neurons and glial cells, <u>Biochem.</u>
     <u>Pharmacol.</u>, 24: 933-938.
Jacobson, M., 1978, "Developmental Neurobiology", Plenum Press,
     New York.
Johnston, G.A.R., and Davies, L.P., 1974, Postnatal changes in
     the high affinity uptake of glycine and GABA in the rat
     central nervous system. <u>J. Neurochem.</u>, 22: 101-105.
Johnston, G.A.R., 1977, Effects of calcium on the potassium-
     stimulated release of radioactive β-alanine and γ-amino-
     butyric acid from slices of rat cerebral cortex and spinal
     cord, <u>Brain Res.</u>, 121: 179-181.

Jóo, F., Dames, W., and Wolff, J.R., 1979, Effect of prolonged
     sodium bromid administration on the fine structure of
     dendrites in the superior ganglion of adult rat. Progr. in
     Brain Res., 51: 109-115.
Katz, R.I., Chase, T.N., and Kopin, I.J., 1969, Effect of ions on
     stimulus-induced release of amino acids from mammalian
     brain slices. J. Neurochem. 16: 961-964.
Levi, G., and Raiteri, M., 1974, Exchange of neurotransmitter amino
     acids at nerve endings can stimulate high-affinity uptake.
     Nature, 250: 735-737.
Martin, D.L., 1976, Carrier-mediated transport and removal of GABA
     from synaptic regions, in: "GABA in Nervous System Function",
     E. Roberts, T.N. Chase, D.B. Tower, eds., Raven Press,
     New York, pp. 347-386.
Meier, E., Drejer, J., and Schousboe, A., 1983, Trophic actions of
     GABA on the development of physiologically active GABA
     receptors, in: "CNS-Receptors from Molecular Pharmacology
     to Behaviour", P. Mandel, F.V. DeFeudis, eds., Raven Press,
     New York, Adv. Biochem. Psychopharmacol. 37: 47-58.
Neal, M.J., and Bowery, N.G., 1977, Cis-3-aminocyclohexane-
     carboxylic acid: a substrate for the neuronal GABA
     transport system, Brain Res., 138: 169-174.
Parnavelas, J.G., and Uylings, H.B.M., 1980, Growth of non-
     pyramidal neurons in the visual cortex of the rat: A
     morphometric study, Brain Res., 193: 373-382.
Raedler, A., and Sievers, J., 1975, The development of the visual
     system of the albino rat, Adv. Anat. Embryol. Cell Biol.,
     50, Springer, Berlin-Heidelberg-New York, pp. 88.
Rakic, P., 1972, Mode of cell migration to the superficial layers
     of fetal monkey neocortex, J. Comp. Neurol., 145: 61-84.
Rakic, P., 1974, Neurons in rhesus monkey visual cortex: systematic
     relation between time of origin and eventual disposition,
     Science, 183: 425-427.
Redburn, D.A., Broome, D., Ferkany, J., and Enna, S.J., 1978,
     Development of rat brain uptake and calcium-dependent
     release of GABA, Brain Res., 152: 511-519.
Ribak, C.E., 1978, Aspinous and sparsely-spinous stellate neurons
     in the visual cortex of rats contain glutamic acid
     decarboxylase, J. Neurocytol., 7: 461-478.
Rickmann, M., Chronwall, B.M., and Wolff, J.R., 1977, On the
     development of non-pyramidal neurons and axons outside the
     cortical plate: The early marginal zone as a pallial anlage,
     Anat. Embryol., 151: 285:307.
Rickmann, M., and Wolff, J.R., 1977, Cytological characteristics
     of early stages of glial differentiatin in the neocortex,
     Folia Morphol. 25: 235-239.
Rickmann, M., and Wolff, J.R., Prenatal Gliogenesis in the neo-
     pallium of rat. Adv. Anat. Embryol. Cell Biol.(in press).

Seiler, N, and Sarhan, S., 1983, Metabolic routes of GABA
    formation in chick embryo brain, Neurochem. Intern.,
    5: 625-633.
Srinivasan, V., Neal, M.J., and Mitchell, J.F., 1969, The effect
    of electrical stimulation and high potassium concen-
    trations on the efflux of [³H]γ-aminobutyric acid from
    brain slices. J. Neurochem., 16: 1235-1244.
Taberner, P.V., Pearce, M.J., and Watkins, J.C., 1977, The
    inhibition of mouse brain glutamate decarboxylase by
    some structural analogues of L-glutamatic acid, Biochem.
    Pharmacol., 26: 345-349.
Weissmann-Nanopolous, D., Belin, M.F., Didier, M., Aguera, M.,
    Partisani, M., Maitre, M., and Pujol, J.F., 1983, Immuno-
    histochemical evidence for neuronal and non-neuronal
    synthesis of GABA in the rat subcommissural organ.
    Neurochem. Intern., 5: 785-791.
Wilson, S.H., Schrier, B.K., Farber, J.L., Thompson, E.J.,
    Rosenberg, R.N., Blume, A.J., and Nirenberg, M.W., 1972,
    Markers for gene expression in cultured cells from the
    nervous system, J. Biol. Chem., Vol. 247, No. 10,
    pp. 3159-3169.
Wolff, J.R., 1978, Ontogenetic aspects of cortical architecture:
    Lamination, in:"Architectonics of the Cerebral Cortex",
    M.A. Brazier and H. Petsche, eds., Raven Press, New York,
    pp. 159-173.
Wolff, J.R., Chronwall, B.M., and Rickmann, M., 1978, Morphogene-
    tic relations between cell migration and synaptogenesis in
    the neocortex of rat, in: "Proceeding of the European
    Society for Neurochemistry", vol. 1., V. Neuhoff, ed.,
    Verlag Chemie, Weinheim-New York, pp. 158-173.
Wolff, J.R., Jóo, F., and Dames, W., 1978, Plasticity in dendrites
    shown by continuous GABA administration in superior
    cervical ganglion of adult rat, Nature, 274: 72-74.
Wolff, J.R., Rickmann, M., and Chronwall, B.M., 1979, Axo-glial
    synapses and GABA-accumulating glial cells in the embryonic
    neocortex of the rat, Cell Tiss. Res., 201: 239-248.
Wolff, J.R., 1981, Some morphogenetic aspects of the development of
    the central nervous system, in: "Behavioral Development.
    The Bielefeld Interdisciplinary Project", K. Immelmann,
    G.W. Barlow, L. Petrinovich and M. Main (eds.), Cambridge
    University Press, New York, pp. 164-190.
Wolff, J.R., 1981, Evidence for a dual role of GABA as a synaptic
    transmitter and a promoter of synaptogenesis, in: "Amino
    Acid Neurotransmitters", F.V. DeFeudis, P. Mandel, eds.,
    Raven Press, New York, pp. 459-466.
Wolff, J.R., Jóo, F., Dames, W., and Fehér, O., 1981, Neuroplastic-
    ity in the superior cervical ganglion as a consequence of
    long-lasting inhibition, in: "Cellular Analogues of Condit-
    ioning and Neural Plasticity", O. Fehér, F. Joó, eds.,
    Adv. Physiol. Sci., Vol. 36, pp. 1-9.

Wolff, J.R., and Chronwall, B.M., 1982, Axosomatic synapses in
     the visual cortex of adult rat. A comparison between GABA-
     accumulating and other neurons, J. Neurocytol., 11:
     409-425.
Wolff, J.R., and Wagner, G.P., 1983, Selforganization in
     synaptogenesis: Interaction between the formation of
     excitatory and inhibitory synapses, in: "Synergetics
     of the Brain", E. Basar, H. Flohr, H. Haken and A.J.
     Mandel, eds., Springer Verlag, Berlin, Heidelberg, New
     York, Tokio, pp. 50-59.
Wolff, J.R., Chronwall, B.M., and Rickmann, M., 1983, "Diffuse
     deposition mode" provides rat visual cortex with non-
     pyramidal and GABA-ergic neurons, 4th Intern. Congr.
     Intern. Soc. for Developm. Neurosci., Abstr. p. 54.
Wolff, J.R., Böttcher, H., Zetzsche, T., Oertel, W.H., and
     Chronwal, B.M., Development of GABA-ergic neurons in rat
     visual cortex as identified by glutamatedecarboxylase-like
     immunoreactivity, Neurosci. Lett. (in press).
Wong, P.T., and McGeer, E., 1981, Postnatal changes of GABA-ergic
     and glutamatergic parameters, Dev. Brain Res. 1:
     519-530.
Wu, J.Y., 1980, Properties of L-glutamate decarboxylase from
     non-neuronal tissues, in: "GABA Neurotransmission",
     H. Lal, ed., Brain Res. Bull., 5, suppl. 2:31-36.

PLASTICITY OF THE DEVELOPING SYNAPSE

Mark C. Fishman

Section on Neurobiology
Developmental Biology Laboratory
Massachusetts General Hospital, Harvard Medical School
Howard Hughes Medical Institute
Boston, MA

"Plasticity" is a term that describes the anatomical, cellular, and molecular reorganizations of the nervous system that occur in response to experience. It serves as a useful rubric to distinguish processes that are environmentally regulated from those that unfold from a rigidly programmed read-out of the genome. Thus, some connections might be termed "plastic" and others "hard-wired". The experience that modifies connections is, of course, ultimately enforced at the molecular level, often through modification of neuronal activity. However, the experimental paradigm may utilize manipulations at a site distant from the actual neurons of interest. Analysis of changes in connections to the cortex during visual deprivation provide one elegant example of the power of this approach[1]. Plasticity of the nervous system is prominent in developing animals, where even transient deprivation may have permanent sequelae.

Figure 1 illustrates the processes of synaptogenesis as functionally divided into categories of "targeting", "stabilization", and "rearrangement". They are pictured as proceding simultaneously, but it is useful to consider them separately, especially during elucidation of their molecular bases. Each of the three stages presumably will have particular molecular tags, or informational molecules. The cellular distribution and amount of some of these molecules may be responsive to the environment, and hence "plastic" in their expression, but it is likely that many will be rigidly programmed in the genome to be expressed in an ordered and invariant progression during development.

241

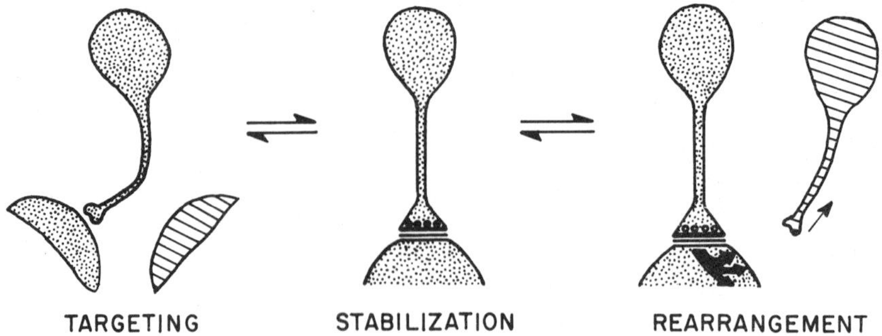

Figure 1. A functional classification of synaptogenesis.

TARGETING

Targeting is a term used to describe the tendency for neurons to
extend processes in a directed fashion, and to selectively form
connections with one cell while bypassing many alternatives. To some
degree it is balanced by an intrinsic tendency for a neuron to
restrict its peripheral field of innervation[2]. This targeting is
fairly accurate, such that when chick spinal cord[3] or amphibian
eyes[4] are rotated early in development, their neuronal projections
innervate the normal target regions, despite having to arrive there
by an abnormal route. Such observations suggested that presynaptic
and postsynaptic cells bear complementary chemical makers. In
vertebrates this targeting does not dictate the exact final pattern
of connections. This is in part because neurons make and receive
many more connections in the perinatal period than they retain in
adulthood. There is a developmental pruning of redundant or
incorrect connections during generation of the adult pattern. In
invertebrates, such as the leech, there appears to be directed growth
of the first growth cone of identified neurons, as well as evidence
for pruning variations[5]. The molecules that guide the directional
growth of the presynaptic cell within the brain's three-dimensional
matrix remain unknown, as do those that specify the class of cells
that are appropriate postsynaptic cells. They may be soluble, or
bound to the matrix or to the cell. It is difficult to envisage that
standing gradients of soluble factors in the extracellular space can
provide more than the most general of directional signals. Even the
apparent chemotactic properties of nerve growth factor on cultured
neurons may not be important _in vivo_[6]. One cell membrane glycoprotein that may carry topographic information has been revealed in
the chick retina by Trisler et al[7].

It seems likely that the remarkable accuracy of targeting
between the 10^{10} neurons of mammalian brain will be found to
require the specific and inflexible expression of particular molecular

codes. These might be expressed seriatum during development. Thus,
there would be one class of cell surface markers to guide a probing
growth cone to a general vicinity and another, or a subset of the
first class, to ensure exact connections with a particular group of
neurons. The date of commitment to this final path might well be the
last cell division. Thus, perhaps the 70,000-100,000 rare copy genes
expressed in the mammalian brain[8] might be specific "connectivity
genes".

STABILIZATION

Some molecules required for synapse stabilization are known.
They account, in part, for the transition from growth cone to stabilized synapse and would include those responsible for transmitter
synthesis, packaging, and secretion in the presynaptic cell and for
receptor synthesis and (at least for some receptors) aggregation at
the synapse in the postsynaptic cell. Synapse-specific molecules
have been described in the vesicle[9,10], presynaptic
membrane[11,12], basal lamina[13], and postsynaptic membrane[14].
Both soluble and membrane-bound factors from the postsynaptic cell
influence the nature and amount of presynaptic transmitter, as well
as the sprouting of the presynaptic terminal[15]. Presynaptic cells
also may release receptor aggregating factors[16].

Components of neurotransmission certainly do exhibit variability
in response to their microenvironment and the amount of neuronal
activity, and thus presumably represent one source of neuronal plasticity. For example, there is an enhancement of tyrosine hydroxylase
activity by soluble factors and neuronal activity[17, 18]. The site
of control of such changes remains unknown, especially with regard to
what proportion occurs at the transcriptional level.

REARRANGEMENT

Not all initial synapses are stabilized. The molecules of such
perinatal rearrangement remain unknown, but the net result is a
reduction in polyneuronal innervation. This pruning serves, presumably, to eliminate the errors of targeting and, perhaps, to provide
some flexibility of assembly in response to the environment.

The method of selection of which synaptic connections to retain
and which to reject is not known. The process is not random, in that
it does not proceed to the point of denervation. It is activity-
dependent, as shown in Fig. 2, with enhanced activity speeding[19,20]
and inactivity slowing[21] the rate of reduction in polyneuronal
innervation. It has been envisaged as a competitive process with, as

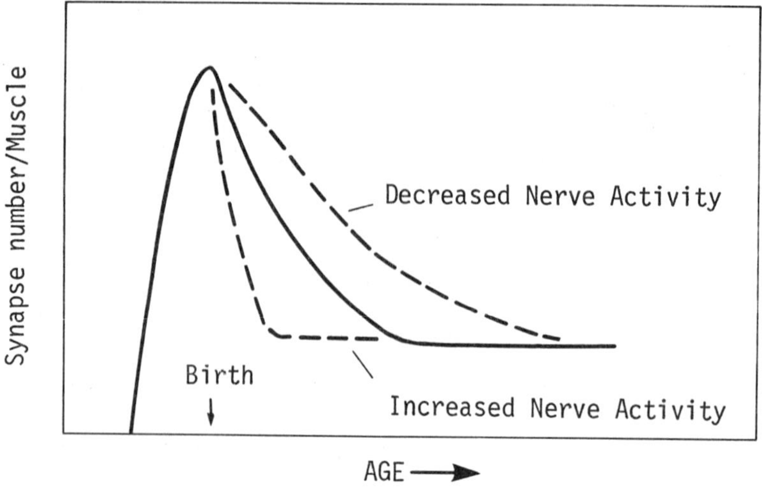

Figure 2.   Synapse elimination at the developing neuromuscular
junction manifests activity dependence.

demonstrated in the visual cortex, active inputs becoming stabilized
at the expense of those that are less active[1]. The absolute size
of innervatable space on the postsynaptic cell membrane seems
limiting, such that a ciliary ganglion cell with six dendrites can
ultimately support more synaptic input than can a ganglion cell with
only one dendrite[22]. The mechanism of transmission failure in
those synapses destined for repression is not known, but, at least at
mammalian neuromuscular junctions, loss of function occurs
abruptly[23], suggesting actual anatomic withdrawal of the repressed
synapse[24].

The plasticity of this phase of synaptogenesis is quite marked,
especially during the "critical period" around the time of birth, and
early deprivation may have permanent sequelae. It seems likely that
control of this "selective stabilization" is epigenetic. The presynaptic cell rejected from one potential target does not collapse
completely in despair, but rather expends its energy enlarging and
improving its synaptic connections with another cell. It seems
possible that such a cell would not need major modifications of gene
expression to accomplish this restructuring, but rather a redirection
of the same gene products to a more restricted peripheral field.

244

REFERENCES

1. Wiesel, T.J. and Hubel, D. (1965). Comparison of the effects of unilateral and bilateral eye closure on cortical unit responses in kittens. J. Neurophysiology. 28: 1029-1040.

2. Brown, M., Jansen, J. and Van Essen, D. (1976). Polyneuronal innervation of skeletal muscle in newborn rats and its elimination during maturation. J. Physiol. (Lond). 26: 387-422.

3. Landmesser, L. (1980). The generation of neuromuscular specificity. Ann. Rev. Neurosci. 3: 279-302.

4. Sperry, R.W. (1943). Visuomotor coordination in the newt (triturus viridescens) after regeneration of the optic nerve. J. Comp. Neurol. 79: 33-55.

5. Kuwada, J. and Kramer A. (1983). Embryonic development of the leech nervous system: primary axon outgrowth of identified neurons. J. Neurosci. 3: 2098-2111.

6. Lumsden, A. and Davies, A. (1983). Earliest sensory nerve fibers are guided to peripheral targets by attractants other than nerve growth factor. Nature. 306: 786-788.

7. Trisler, G.D., Schneider, M. and Nirenberg, M. (1981). A topographic gradient of molecules in retina can be used to identify neuron position. Proc. Natl. Acad. Sci. 78: 2145-2149.

8. Chaudhari, N. and Hahn, W.E. (1983). Genetic expression in the developing brain. Science. 220: 924-928.

9. Matthew, W.D., Tsavaler, L. and Reichardt, L.F. (1981). Identification of a synaptic vesicle-specific membrane protein with a wide distribution in neuronal and neurosecretory tissue. J. Cell Biol. 91: 257-269.

10. DeCamilli, P., Harris, S., Huttner, W. and Greengard, P. (1983). Synapsin 1 (Protein 1), a nerve-terminal-specific phosphoprotein II: its specific association with synaptic vesicles demonstrated by immunocytochemistry in agarose-embedded synaptosomes. J. Cell Biol. 96: 1355-1373.

11. Morel, N., Manaranche, R., Israel, M. and Gulik-Krzywicki. (1982). Isolation of a presynaptic plasma membrane fraction from Torpedo cholinergic synaptosomes: evidence for a specific protein. J. Cell Biol. 93: 349-356.

12. Mijanich, G.P., Porasier, A.R. and Kelly, R.B. (1982). Partial purification of presynaptic plasma membrane by immunoabsorption. J. Cell Biol. 94: 88-96.

13. Sanes, J.R. (1983). Roles of extracellular matrix in neural development. Ann. Rev. Physiol. 45: 581-600.

14. Burden, S. (1981). Monoclonal antibodies to the frog nerve-muscle synapse. In - Monoclonal Antibodies to Neural Antigens (eds - McKay, R., Raff, M., Reichardt, L.F.) p. 247-257. Cold Spring Harbor.

15. Henderson, C., Huchet, M. and Changeux, J.P. (1983). Denervation
     increases a neurite-promoting activity in extracts of
     skeletal muscle. Nature. 302: 609-611.
16. Christian, C.N., Daniels, M.P., Sugiyama, H., Vogel, Z.,
     Jacques, L. and Nelson, P.G. (1978). A factor from neurons
     increases the number of acetylcholine receptor aggregates in
     cultured muscle cells. Proc. Natl. Acad. Sci., USA. 75:
     4011-4015.
17. Ip, N., Perlman, R. and Zigmond, R.E. (1983). Acute trans-
     synaptic regulation of tyrosine 3-monooxygenase activity in
     the rat superior cervical ganglion: evidence for both
     cholinergic and noncholinergic mechanisms. Proc. Natl. Acad.
     Sci., USA. 80: 2081-2085.
18. Patterson, P.H. and Chun, L.L.Y. (1974). The influence of
     non-neuronal cells on catecholamine and acetylcholine syn-
     thesis and accumulation in cultures of dissociated sympa-
     thetic neurons. Proc. Natl. Acad. Sci., USA. 71: 3607-3610.
19. O'Brien, R., Ostberg and A., Vrbova. (1978). Observations on
     the elimination of polyneuronal innervation in developing
     mammalian skeletal muscle. J. Physiol. 282: 571-582.
20. Fishman, M.C. and Nelson, P.G. (1981). Depolarization-induced
     synaptic plasticity at cholinergic synapses in tissue
     culture. J. Neurosci. 1: 1043-1051.
21. Thompson, W., Kuffler, D. and Jansen, J. (1979). The effect of
     prolonged, reversible block of nerve impulses on the elimina-
     tion of polyneural innervation of newborn rat skeletal muscle
     fibers. Neuroscience. 4: 271-281.
22. Purves, D. and Hume, R. (1981). The relation of postsynaptic
     geometry to the number of presynaptic axons that innervate
     autonomic ganglion cells. J. Neurosci. 1: 441-452.
23. Miyata, Y. and Yoskioka. (1980). Selective elimination of motor
     nerve terminals in the rat soleus muscle during development.
     J. Physiol. (Lond). 309: 631-646.
24. Korneliussen, H. and Jansen, J. (1976). Morphological aspects
     of the elimination of polyneuronal innervation of skeletal
     muscle fibers in newborn rats. J. Neurocytol. 5: 591-604.

ACTIVITY AND COMPETITION-DEPENDENT

SYNAPSE REPRESSION IN CULTURE

Mark C. Fishman and Phillip G. Nelson

Section on Neurobiology
Developmental Biology Laboratory
Massachusetts General Hospital and Harvard Medical School
and Howard Hughes Medical Institute
Boston, MA
and Laboratory of Developmental Neurobiology
National Institutes of Health
Bethesda, MD

Throughout the nervous system neurons receive many more synaptic inputs prior to birth than are retained into adulthood[1]. Regulators of this rearrangement are poorly understood, but seem to include an element of competition between inputs for particular domains of space on a postsynaptic cell. The competition depends in part upon the amount of neuronal activity or its patterning, and in part upon ill-defined components of the pre- and postsynaptic cells that make some connections "appropriate" and some foreign, or "inappropriate". Were such rearrangements to occur in cell culture systems they might be more accessible to cellular, biochemical, and molecular genetic characterization. In the restricted environment of cell culture many options for the cells are better controlled, especially when homogeneous populations are used, and electrophysiological access for assay of synapse formation is straightforward. Synapse plasticity can be examined when manipulating the cellular and soluble constituents as well as the degree of neuronal activity.

COMPETITION BETWEEN CHOLINERGIC NEURONS

Neurons of the chick ciliary ganglion, which normally innervate the muscles of the iris, choroid and ciliary body of the eye, form cholinergic synapses with thoracic skeletal myotubes in culture[2]. We examined: (1) whether such "inappropriate" innervation could be eliminated by competition with "appropriate" innervation from spinal motoneurons, and (2) whether activity played a role in controlling the establishment and rearrangement of these synapses[3].

Ciliary ganglia explants were plated onto a low density culture
of chick myotubes. The ganglia extended a halo of processes which
covered neighboring myotubes. Synapses were assessed starting three
days after the plating of the ganglion by intracellular recording
from myotubes during extracellular stimulation of the ganglion and a
myotube counted as connected if synaptic potentials were elicited in
it by ganglion stimulation. As at the immature synapse, quantal
efficiency was low and fatiguability rapid. Most myotubes in the
vicinity of the ganglion (one field diameter - about 500 um) were, in
fact, synaptically connected to it. This was true whether the fiber
network covering the myotube was a dense plexus or a few scattered
fibers. Polyneuronal innervation of a given myotube was not
uncommon, as revealed by using incremental increases in stimulating
potential to exceed the threshold of different neurons. However, the
degree of polyneuronal innervation was difficult to quantitate
because of fatiguability of transmission, so each myotube was scored
as either connected or not. Connections were assessed from three to
six days after plating. The number of myotubes that manifested
synaptic input diminished only slightly over this time, from 96% to
81%.

Experiments were performed by plating dissociated spinal cord
cells on the myotubes prior to the addition of the ganglia. As shown
in Figure 1, the presence of these cells markedly affected the input
from the ganglia. Under these conditions, as shown in Figure 1, the
initial innervation by the ciliary ganglion was relatively
unaffected, so that about 80% of myotubes were connected when first
assayed. However, during subsequent days, input from the ciliary
ganglion was reduced to 17%.

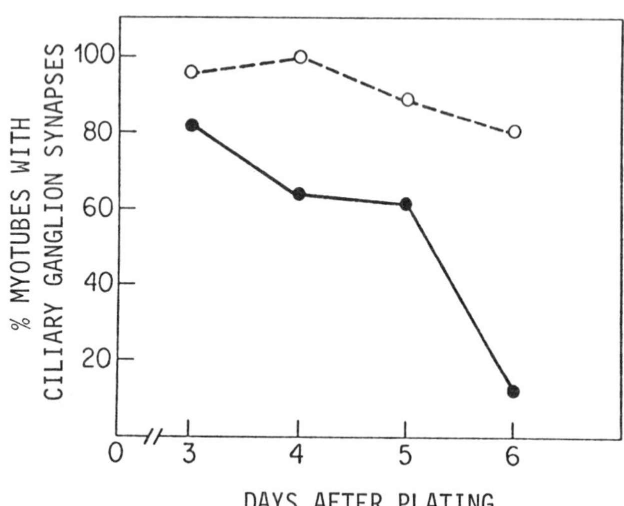

Figure 1. The percentage of myotubes connected to the ciliary
ganglion in the absence (o) and presence (●) of competing
spinal cord cells. (From Schaffner et al., 1984).

248

This suggested that spinal cord cells did not block the initial
formation of synapses, but rather reduced their stability. We have
termed this phenomenon "repression". It also occurred when spinal
cord neurons were added after the ciliary ganglion had already formed
connections[4], again suggesting an interference with stability of
synapses rather than with their formation. Repression exhibited cell
type specificity. Thus, neurons other than spinal motoneurons (from
dorsal spinal cord, dorsal root ganglia, or cortex) did not exert
this competitive effect. Medium conditioned by motoneurons and
myotubes similarly was without repressive actions. This suggested
that spinal cord neurons exerted the repression by direct cell-cell
contact. Another possibility is that the responsible molecules might
be soluble but very labile and active only over short distances or
brief durations.

ACTIVITY AND SYNAPSE REPRESSION

     The role of activity in the spinal cord cells' repression of
ciliary ganglion-myotube synapses was assessed using curare to
interrupt neuromuscular transmission. Curare, in concentrations
sufficient to block neuromuscular transmission, did not prevent the
initial formation of synapses between ciliary ganglion and myotubes.
However, it did partially reverse the repression caused by spinal
cord cells, as shown in Figure 2. About one-third of the repressive
effect of spinal cord cells was activity-dependent, by this assay.
However, two-thirds of the effect of spinal cord cells was therefore
activity-independent.

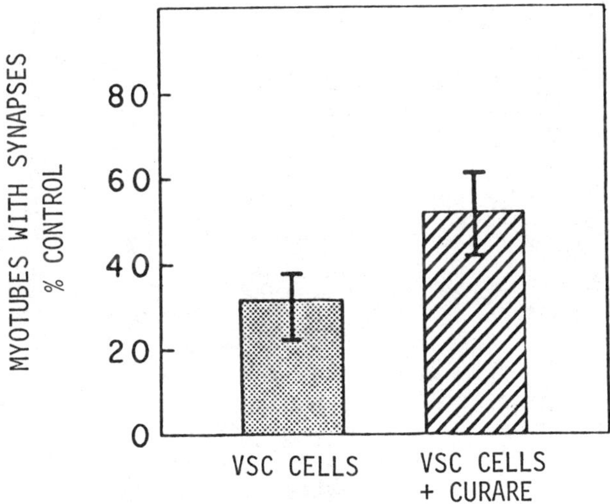

Figure 2. Curare partially reverses the repression of ciliary
           ganglion-myotube synapes caused by ventral spinal cord
           cells (VSC) cells. (From Schaffner et al., 1984).

In order to evaluate the role of activity in synaptogenesis more directly, we studied the cholinergic synapse between the clonally derived hybrid cell NG108-15 (neuroblastoma X glioma) and myotubes[5]. These cells form synapses that morphologically and physiologically resemble immature synapses, with a paucity of vesicles and synaptic membrane specializations, low quantal efficiency, and rapid fatiguability of transmission. The system has advantages for studies of activity-dependence in that (1): spontaneous spike activity is negligible so that membrane potential can be modulated without attendant complicating changes in spike activity, and (2): NG108-15 cells do not form synapses with one another, so that unlike most other synapse-forming cell culture systems, cells are either presynaptic (NG108-15 cells) or post-synaptic (myotubes), but not both. The cells are large, so that connections between particular cells can be assessed by intracellular recording and epp analysis, thus quantitating both spontaneous and evoked transmitter release.

NG108-15 cells contact myotubes both via their soma and their processes. In almost all cases where such contact was visible, evidence of synaptic connection could be obtained. As at the immature neuromuscular junction, polyneuronal innervation of one myotube was prominent.

The paradigm used for chronic stimulation was veratridine-induced depolarization. The membrane potential of NG108-15 cells diminished by 20mV in the veratridine. Prior to testing, the veratridine was removed. After 24 hours of depolarization, the number of connections from NG108-15 cells was markedly diminished, to about 30% of matched controls. The pattern of synapse repression was not random, but seemed to reflect a diminution in polyneuronal inner-vation. Thus, repression did not proceed to the point of denerva-tion. However, the number of myotubes that could be documented to have received more than one input, by stimulating several NG108-15 cells contacting the same myotube, diminished from 50% to about 6%. The remaining synapses were unaltered in their efficacy, as revealed by quantal content measurement. So, as occurs _in vivo_, synapse repression occurred abruptly, and was not preceeded by a period of dysfunction.

Taken together these series of experiments suggest that synapse rearrangement can occur in cell culture secondary to physiological manipulation. There is competition between presynaptic cells based on cell type specificity and neuronal activity. Presynaptic cells that are more "appropriate" with regard to a particular target cell seem to have an advantage in the competition, a process that appears to require direct cell-cell contact. One model might be that synapse formation is a dynamic process, with rates of formation and elimina-tion determining the final number of synapses at equilibrium. The rate of elimination might be enhanced by increased activity and the

presence of more "appropriate" competing presynaptic cells. The
process of elimination, however, ceases in the absence of competing
cells and when the system is silenced. The net result is a reduction
in the degree of polyneuronal innervation, but never to the point of
denervation.

The molecular basis of such repression remains to be determined. In part it may reflect changes whereby the postsynaptic cell
limits the domains of innervatability, for example, by clustering
receptors. In part it may reflect presynaptic alterations that
reflect the transition from growth cone to stabilized synapse. In
order to evaluate these possibilities we designed systems to compare
the proteins characteristic of growth cones and synapses.

PRESYNAPTIC PROTEINS DURING SYNAPTOGENESIS

Dorsal root ganglion (DRG) neurons form synapses with spinal
cord neurons in culture. In order to identify presynaptic proteins
that change during synaptogenesis we designed the system shown in
Figure 3[6]. This system allowed selective metabolic labelling of
proteins in the DRG cells that were transported to the processes of
the cells. DRG cells were plated in the central compartment. Their
cell bodies were restricted to that chamber, and grease seals pre-
vented bulk flow between the chambers. DRG processes, however,
extended under the seals into the side chambers. Metabolic labelling
with [^{35}S]-methionine and subsequent two-dimensional electrophoresis of the proteins from the side chambers, thus provided a
profile of proteins in the growing axons of the DRG cells. This
profile included more than 300 proteins and was characteristic for
different populations of neurons[7]. When dorsal root ganglion cells
were plated in the central slit, their processes would extend under
the barrier and innervate spinal cord cells in the side compartment,
as revealed by electrical evidence of synaptic input.

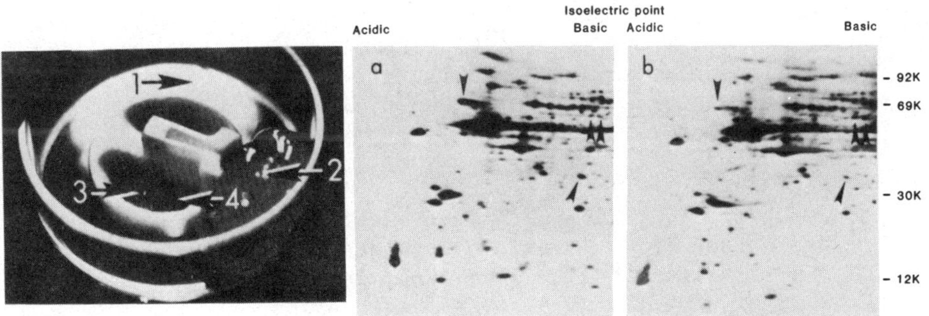

Figure 3. The chamber used to study axonally transported proteins,
and a comparison of the two-dimensional gel pattern of
[^{35}S] methionine-labelled proteins in DRG axons in the
absence (a) and presence (b) of postsynaptic target
neurons. (From Sonderegger et al., 1983).

Concommitant with the formation of synapses, four axonally transported DRG proteins changed in their expression, as shown in
Figure 3. Two proteins were reduced in amount, one with apparent
molecular weight (M.W.) of 65,000 and one with a M.W. of about
50,000. Two proteins with M.W. of about 60,000 became more
abundant. As for some of the synaptic interactions described above,
cell contact seemed an important component in causing the changes,
and spinal cord conditioned medium did not exert the effects.

This system may provide a handle on several of the more abundant
proteins that are associated with axon elongation and synapse formation. For example, the 65K protein has a molecular weight and
isoelectric point similar to a subunit of the neurofilament protein.
Its diminution during synapse formation may reflect cessation of axon
elongation. The 60K proteins whose concentrations are increased may
relate to synapse stabilization.

It will likely be necessary to turn to the powerful tools of
cDNA cloning and differential hybridization in order to reveal
changes in gene expression of rarer mRNA's than can be resolved by
these protein gels. We are at present trying to utilize this
combination of cell biology, electrophysiology, protein chemistry and
molecular genetics to identify other molecules of synaptogenesis.

REFERENCES

1. Fishman, M.C. (1984). This Volume.
2. Betz, W. (1976). The formation of synapses between chick embryo
      skeletal muscle and ciliary ganglia grown In Vitro. Journal
      of Physiology, 254: 63-73.
3. Schaffner, A.E., Nelson, P.G. and Fishman, M.C. (1984). Synapse
      repression in cell culture, Dev. Brain Res. (In Press).
4. Fishman, M.C., Schaffner, A.E. and Nelson, P.G. (1982). Synapse
      repression in culture, Soc. for Neuroscience Abst. 8: 641.
5. Fishman, M.C. and Nelson, P.G. (1981). Depolarization-induced
      synaptic plasticity at cholinergic synapses in tissue
      culture. Journal of Neuroscience, 1: 1043-1051
6. Sonderegger, P., Fishman, M.C., Bokum, M., Bauer, H.
      and Nelson, P.G. (1983). Axonal proteins of presynaptic
      neurons change during synaptogenesis, Science, 221: 1294-1296.
7. Sonderegger, P., Fishman, M.C., Bokum, M., Bauer, H.
      and Nelson, P.G. (1984). A few axonal proteins distinguish
      ventral spinal cord neurons from dorsal root ganglion neurons,
      Journal of Cellular Biology, 98: 364-368.